Oil in West Texas and New Mexico

Phillips Petroleum Company's rig No. 20 near Fort Stockton in the early 1960's. *PBPM*.

Oil in West Texas and New Mexico

A Pictorial History of the Permian Basin

By WALTER RUNDELL, JR.

Published for the PERMIAN BASIN PETROLEUM MUSEUM,
Library, and Hall of Fame, Midland, Texas
By TEXAS A&M UNIVERSITY PRESS, COLLEGE STATION

Library of Congress Cataloging in Publication Data

Rundell, Walter.
    Oil in West Texas and New Mexico.

    Bibliography: p.
    Includes index.
    1. Petroleum industry and trade—Texas—History.
2. Petroleum industry and trade—New Mexico—History.
I. Title.
TN 872.T4R86   338.2′7282′097648   81-48376
ISBN 0-89096-125-5   AACR2

Manufactured in the United States of America

FIRST EDITION

For Jack L. Young

Contents

Preface

ANY historical account, whether literary or visual, is shaped by available sources. This history of oil in the Permian Basin relies largely on photographs but also includes other kinds of illustrations. Photographs have documented the petroleum industry in West Texas and New Mexico extensively, yet the nature of the industry has always dictated some imbalances. Since production is the most exciting phase of the business, photographers have concentrated on that. Likewise, many pictures of refineries exist, lending a sense of time and place but showing little more than plant exteriors. Since refining operations are largely internal, they elude the photographer. Conversely, pictures of drilling equipment reveal a great deal about the technology of production. The chronology of petroleum in the Permian Basin has similarly influenced the photographic record—or at least that part available for research. Archivists and collectors have covered the early days of the industry better than the later, so much fuller documentation exists for the discoveries of the 1920's than of the 1960's—then too, there was simply more activity in the early days. As a consequence of these facts, this book incorporates what photographers have thought important, and I have striven through selection and organization to present a representative account.

The organization is chronological by county, beginning with the initial discovery in Mitchell County in 1920. Commercial production began at later dates in other counties, and the book deals with each county in turn. Thus, within the chronological framework of discoveries, the book proceeds primarily on a county basis, rather than field-by-field. Any form of historical organization has its artificialities and problems, this included. But since the development of oil fields directly affected county governments, especially county seats, and was inextricably linked to the town-building process, this form of organization serves best, both in text and pictures.

The last parts of the book are not tied to the regional organization, since their topics pertain to the entire area. These include the scientific aspects of exploration, including the work of geologists and seismologists, as well as industrial technology. Since the work of pipeliners, particularly important in transporting Permian Basin oil and gas to population centers for refining, is buried underground, visual documentation of the process of laying pipe is especially valuable. Altogether, this pictorial record of oil, from the geologist's initial calculations to gasoline in the service station pump, testifies to an industry that has vitalized the vast Permian Basin.

The creation of this pictorial record was as predictable as the appearance of the boom towns that accompanied major discoveries of oil. By their nature, oil towns have always attracted those seeking economic gain. They have been a magnet for skilled and unskilled workers in the fields, for promoters, developers, merchants, professional people, and others representing the accoutrements of civilization. The lure of money has been strong enough to draw people into desert wastelands, and with sufficient economic motive people have turned such locales into viable communities. Among those attracted to oil towns have been commercial photographers. They came seeking a livelihood by creating a pictorial record of the people and events of the region. Some photographers relied upon people's instinct to preserve personal histories through individual portraits and family pictures and thus limited themselves to studio work. Others hoped that those participating in the historical events connected with the petroleum business would want some visual record and took pictures all round the Permian Basin.

To these venturesome photographers we owe a great debt, since they have left an account that vivifies the written sources of history. Some doubtless prospered, while others did not; irrespective of commercial success, their contribution to historical understanding was in direct ratio to their skill with the camera. This skill is particularly impressive considering the cumbersome equipment photographers had to take into the field in the 1920's and 1930's. For instance, that he had his camera prepared to capture the explosion of Skelly-Amerada University No. 1 was a considerable tribute to the talent of the Permian Basin's leading photographer, Jack Nolan. His legacy, along with that of others, is immense. Among others who made notable contributions to the region's photographic record were Adams Studio, E. J. Banks, Doubleday, and Bill Shoopman. Besides the legacy left by commercial photographers, oil companies themselves realized the importance of creating a photographic record of their activities and have thus added to the documentation. In addition to these professionals, many amateur photographers have enriched the record with fine snapshots. With these collective contributions, the history of most aspects of the Permian Basin oil industry is documented visually.

Acknowledgments

As essential as was the work of photographers to this book, equally important was that of the repositories that have gathered and preserved their pictures. From the staffs of these repositories I have received the most cordial cooperation, and it gives pleasure to offer public appreciation. Without the collections of the Permian Basin Petroleum Museum, Library and Hall of Fame, this book would have been impossible; therefore, to its director, Homer T. Fort, and archivist, Betty J. Orbeck, I express gratitude for their unfailing assistance. Because my research there involved thirty-three collections, in the book I have attributed pictures to the collections they came from, plus the initials PBPM to indicate the repository. Those photographs from the museum's own collection bear only the PBPM identification. I list only the names of other repositories, since the number of collections researched in them was limited. The bibliography gives full information about individual collections.

These other institutions and their helpful staff members are as follows: American Petroleum Institute, Richard E. Drew; Institute of Texan Cultures, Tom F. Shelton; Mahan and Associates, Inc., Richard L. Mahan and Linda Mahan; Odessa *American*, Bob Horn; Texaco, Inc., Kenneth E. McCullam and Stafford Acher; Texas Mid-Continent Oil and Gas Association, John J. Cassel and Jack Rolf; Texas State Archives, Jean Carefoot; Texas Tech University's Southwest Collection, David J. Murrah and Michael Q. Hooks; University of Texas at Austin Archives, Jeanne R. Willson and Ralph L. Elder; University of Texas at Austin Humanities Research Center, Mrs. May Ellen MacNamara; and University of Texas of the Permian Basin's Permian Historical Society Archives, Mrs. Bobbie Jean Klepper.

The American Petroleum Institute and Texas Mid-Continent Oil and Gas Association have performed a valuable service by collecting historical photographs. Most of their photographs have come from corporations, and each repository requests that the name of the donor be credited. For those corporations which still exist under the original name, I have honored that request. Otherwise, I cite the repository.

Almost as essential to this book as the repositories that have collected pictures has been the two-volume work of Samuel D. Myres, *The Permian Basin: Petroleum Empire of the Southwest.* His encyclopedic books furnish the starting point for any further serious history of the region. I also wish to acknowledge further the expert photographic services of Mahan and Associates, Inc., of Odessa. Richard Mahan's interest and involvement in this book have been most gratifying.

Many have graciously given permission for inclusion of works of their own creation or to which they hold the copyright. Although they are credited in captions, I extend special thanks to the Abell-Hanger Foundation, George W. Bush, José Cisneros, Robert A. Estes, Lee Jones, Jr., Mahan and Associates, Inc., Peter Flagg Maxson, and Mrs. Jimmie Moore. In June, 1978, the Abell-Hanger Foundation deposited the Samuel D. Myres Collection of papers, photographs, and maps in the Permian Basin Petroleum Museum, Library and Hall of Fame to be administered by the institution's archivist. The benefactions of the Abell-Hanger Foundation reflect the historical interests of the late George T. Abell and his wife, Gladys Hanger Abell. Abell was the moving spirit behind the Permian Basin Petroleum Museum, Library and Hall of Fame, as well as Myres's two-volume study. His concern for preserving and publishing the historical record of the region lives on in this book, as well, and I wish to pay tribute to the memory of one who believed that understanding the past should lead to enlightened choices for the future.

My sincere appreciation goes to those benefactors of the museum who made this book possible, Mr. and Mrs. Ford Chapman and Mr. and Mrs. Richard Donnelly. Their active interest in history is testified to by the collections they have donated to the PBPM, as well as its publication program. Donnelly's assistance extended to flying tours that enabled me to grasp the enormity of petroleum production throughout the Permian Basin (one flight graciously piloted by Richard Donnelly, Jr.) and taking photographs of drilling operations in July, 1980.

Several individuals have helped in pinning down specific historical facts, and I wish to thank Vicky Garman, Earl David, and Don Zeller of the public relations staffs of Shell, Phillips, and Indiana Standard, respectively; David Doan, my colleague in geology at the University of Maryland; Robert L. Hardesty, Vice Chancellor for Administration, University of Texas System; and particularly John P. Hammett, a petroleum engineer with Arco Oil and Gas in Midland, who taught me the historical value of well tickets.

Special gratitude goes to my family: to David, who helped with the driving on the research trip; to Shelley, who assisted with proofreading; to Jennifer, who provided a diverting counterpoint to the work on this book; and to Deanna, whose critical judgment and assistance have been invariably valuable.

For the cooperation of all who have assisted in supplying information for this book, I am most grateful. The responsibility for the contents is mine alone.

Walter Rundell, Jr.

University of Maryland
January, 1981

Oil in West Texas and New Mexico

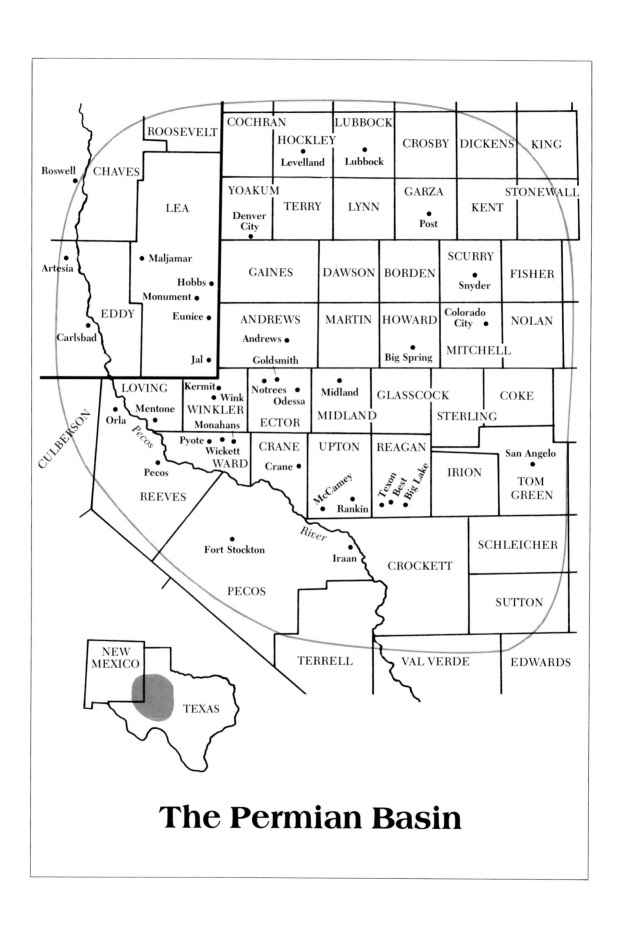

The Permian Basin

The Permian Basin—
Petroleum and People

THE Permian Basin of West Texas and the southeast corner of New Mexico is one of the few areas in the United States whose geological designation has become common geographical usage. The name derives from the city and province of Perm, west of the Ural Mountains in the Soviet Union. In 1841 British geologist Sir Roderick I. Murchison first identified distinctive rock formations in that area. Other places in the earth where such sedimentary beds occur have likewise received the designation of Permian, since they were all formed during that geological age. Of the world's Permian formations, the most notable from the standpoint of petroleum production has been the approximately 68,000-square-mile area in West Texas and New Mexico. The producing area of the Permian Basin is almost square, measuring about 260 miles on each axis. The Texas portion of the Basin extends from Lubbock County and its neighbors on the north to Crockett County on the south. The east-west boundaries go from Tom Green to Culberson County. The New Mexico section of the Basin consists of Lea County and portions of Eddy, Chaves, and Roosevelt counties. The non-petroleum-producing areas of the Basin extend farther to the west and southwest—about to the western edges of Eddy, Culberson, and Jeff Davis counties and encompassing parts of Presidio and Brewster counties.

In aeons past, approximately 200 million years ago, a salt sea covered the area. As the structure of the earth formed and changed, a limestone floor developed in the sea, to be overlaid by other types of rocks, such as dolomite, anhydrite, and more salt. When rivers emptied sand and red clay into the sea, they covered the plants and animals whose remains yielded the hydrocarbons that later resulted in the formation of oil and gas. Geological phenomena of particular importance were the upheavals that created vast mountain ranges beneath and above the surface. During the development of these ranges, layers of rock shifted, trapping deposits of hydrocarbons in faults, anticlines, and salt domes. Another important geological feature was the formation of giant coral reefs. As the millenia deposited their marine life—vegetable and animal—on the reefs, that life furnished the basis for the formation of petroleum.

While surface characteristics of the Permian Basin vary considerably, the entire area is deficient in rainfall, ranging from the semiarid eastern portions, where the average annual precipitation is just short of twenty inches, to the arid western section, where it is just above ten inches. The topography ranges from the high plains of its northern borders—called the Llano Estacado, or Staked Plains, by Spanish explorers—to the mountainous terrain of the Yates field in Pecos County. Because of its physical conditions, the Permian Basin has supported little farming. Where irrigation has been available from wells and the few creeks and rivers in the area, limited cultivation has been possible. One of the striking aspects of the region is the almost unbelievable difference just a little water can make. The Pecos River, for instance, runs like a life-giving artery through the desert expanses of New Mexico and Texas. Its water greens a few acres on each side, enabling them to produce bounteous crops. This irrigation demonstrates the fertility of the land, which without water remains sterile. Observations from the air best confirm the transforming power of water on arid soil.

Because of the nature of the region, the only way people made a living there before the discovery of petroleum was by ranching. Since vegetation was sparse, ranches had to be huge to form economically viable units. Loving County, for example, was so arid that a section (640 acres) would support only four or five cattle. In the post–Civil War era, several notable men undertook ranching in the region, among them C. C. Slaughter, whose Lazy S ranch occupied land in Borden, Dawson, Howard, and Martin counties; John S. Chisum, who located in Chaves County; John T. McElroy, who operated in Pecos, Crane, Upton, Reeves, and Loving counties; the Cowden brothers, George G. and William H., who with their sons ranched in Ward and Lea counties; and John Scharbauer, who raised cattle in Midland and Ector counties. In the early days men ran their cattle on public lands, where the grazing was free. As they accumulated some cash, they began to buy the land. A Texas law of 1874 set the price at $1.50 per acre, but an 1883 law raised the price to $2.00 an acre for land without water and $3.00 for land with. The same law enabled ranchers to lease public land for four cents an acre up to ten years. Under these provisions C. C. Slaughter gained control of over two million acres in West Texas. Although cattle predominated in the area, some ranchers raised sheep. Scharbauer, for instance, began with sheep but switched to cattle in 1888. In 1885 J. M. Shannon settled in Crockett County with a flock of sheep and became a leading citizen of the region.

In addition to ranching, railroads provided about the only other economic activity in the area. Although the Butterfield Overland Mail's famous ox-bow route ran through West Texas and southeastern New Mexico in 1858, the first permanent transportation in the area came with the building of the Texas and Pacific Railway in 1881. This line extended from Fort Worth to El Paso, and as it progressed companies and individuals bought land and hired surveyors to plat townsites. The surveyors selected townsites because of the availability of water for the locomotives' steam engines. Owners of the railroad also hoped that towns would develop to provide way traffic, as well as a market for goods shipped in. Among the towns created as a result of the Texas and Pacific were Colorado City (originally called Colorado), Big Spring, Midland (originally Midway, lying equidistant between Fort Worth and El Paso), Odessa, Monahans, and Pecos. The railroad performed a twofold function in helping develop the early economy of the region: it hauled supplies in, and it enabled ranchers to ship their animals out, which proved much more economical than trail drives.

The State of Texas, eager to promote a rail connection to El Paso, deeded 5,338,528 acres of public domain to the Texas and Pacific Railway. This land, which included the right-of-way, gave the corporation the resources to finance construction, and the railroad raised cash by selling land to individuals. It administered the bulk of the land through the Texas Pacific Land Trust. Many oil wells were located on railroad acreage, most notably the Permian Basin's discovery well, T. and P.–Abrams No. 1 in Mitchell County. W. H. Abrams was the agent for the railroad's trust. The trust land also produced oil in Reagan, Glasscock, Midland, Ector, Reeves, and Culberson counties. The TXL field in Ector County was an exceedingly rich discovery on railroad lands.

Another link in the transportation system of the region came with the construction of the Pecos Valley Railway in 1890–1891. The line, following the river, connected Pecos with Carlsbad, Artesia, and Roswell. Of even greater importance was the Kansas City, Mexico, and Orient Railway, promoted by Arthur E. Stilwell, who dreamed of connecting the Missouri River Valley with Topolobampo, a Mexican port on the Pacific, where goods could be shipped directly to the Orient. The road came through Kansas and Oklahoma, and in the early years of this century headed toward Sweetwater and San Angelo. When the Orient railroad transferred its general offices and machine shops from Sweetwater to San Angelo shortly after 1910, the former obtained an injunction to retain the facilities, but the state supreme court supported San Angelo. From San Angelo the road ran through Big Lake, Rankin, Fort Stockton, and Alpine. Because of the connections west from San Angelo, the Orient hauled much of the freight for developing the Big Lake and McCamey fields. When the Orient went into receivership in 1928, the Santa Fe took over.

Before the discovery of oil in the Permian Basin in 1920, the region had all the primary characteristics of America's western frontier. The population was sparse because the vast stretches of semiarid and desert lands could support only limited ranching and no intensive cultivation. The towns in the region were county seats and those which developed in connection with the railroads. In many cases, these were one and the same, such as Fort Stockton, Pecos, Odessa, Midland, Big Spring, and Colorado City. Pioneers who settled the region possessed the characteristics ascribed to frontiersmen by Frederick Jackson Turner: self-reliance, inventiveness, optimism, and individualism. A harsh nature forced them to deal with life realistically. Those who lived by ranching understood that theirs was an economic gamble, since droughts occurred far more often than adequate rainfall. When droughts killed native grasses, ranchers had to import expensive feed, since they could grow little for themselves, or else let their herds starve.

The same droughts that diminished grazing also dried up water supplies. Of course, not every year brought natural disasters to ranchers, but they had to make big profits on occasion to tide them over the lean years. The hope for such profits created the optimism that enabled them to endure the hard physical conditions and demands of their frontier.

Only the hardy and self-reliant ventured into the area originally, for the terrain and weather discouraged the timid. But this frontier also acted as a crucible for those who came to prove themselves. Men and women who withstood the privations of life where water was scarce, where human contacts were limited, and where work was unremittingly hard developed individualistic outlooks. They knew that they had the responsibility for their own existence and welfare—that no one else was going to ride out to doctor their sick calf on a wintry night or to protect their sheep from coyotes. Those who could not withstand the privations, demands, and challenges of the region quickly retreated to where the physical demands of life were less drastic—usually eastward, where green grass and trees denoted a benign nature and where towns and cities offered more amenities of civilization.

As Richard Wade noted in his influential book *The Urban Frontier*, towns often served as outposts for the agricultural frontier. They did not develop in the aftermath of settling, providing a commercial and social center for a maturing farming economy. Rather, they served as spearheads of civilization, offering pioneers the means for settlement. The towns' seed stores, banks, farm implement stores, churches, and schools offered the institutional means for the westward movement. Of course, Wade's thesis applied to the trans-Appalachian frontier, but as American civilization crossed the Mississippi, the same conditions generally prevailed.

Certainly, there was little Anglo-American settlement in the Permian Basin before the Texas and Pacific Railway came in 1881. Prior to that, many counties existed in name only, since their areas had little or no population. In such cases, the unorganized counties were administered from neighboring county seats. For instance, Upton County, created in 1887, remained unorganized until 1910; Crane County, dating from 1887, became organized only in 1927. When the state legislature created Tom Green County in 1874, it included most of the Permian Basin. Two years later the legislature carved up Tom Green into fifty-four additional counties. Then in the mid-1880's the western part of the remaining county was divided into six counties, including Midland and Ector. Midland became organized in 1885 and Ector, in 1891. Loving County, named for Oliver Loving since the cattle trail bearing his and Charles Goodnight's names followed the Pecos River, which formed the county's western border, was legally organized in 1931, having previously been attached to Reeves County for tax and judicial purposes. When the railroad spawned towns that could furnish water for their steam locomotives, it provided the nuclei for settlement. From the seeds planted by the railroads that laid track through the region, towns grew to nurture the ranching economy and then, after 1920, the oil boom. Whereas the towns and ranches prior to 1920 had been unspectacular in economic growth, after that date petroleum vitally altered both the urban centers and the region's economy.

Most of the oil fields in the Permian Basin led either to the founding of new towns or to the dramatic development of nearby communities. The quantity of oil produced and the geographical location seem to have been the major determinants in town building. Any time a major oil field developed away from an established community, a town would spring up. If the oil discovery was modest, the nearest town would usually serve the needs of the producing area. In those instances in which production was near a town, the oil field usually turned the town into a city, as it became a center for supplies, transportation, management, and financing.

Underwriters' T. and P.-Abrams No. 1, discovery well in the Permian Basin, was spudded February 8, 1920, and began producing in late June. The well is near Westbrook in Mitchell County. Drillers Wilbur J. Thomson and Dan Lewis stand on the derrick floor. Extending from the right side of the derrick are wrench poles, counterweights that enable drillers to lift the heavy wrenches used to set up and unscrew cable tool drilling joints. When Underwriters Producing and Refining Company sold the discovery well to American Petrofina Company of Texas, the latter renamed it Westbrook No. 701. For secondary recovery, Petrofina has water-flooded the field, and No. 701 now produces about two barrels per day. *Lee Jones, Jr.*

The Discovery: Mitchell County

THE discovery of oil in the Permian Basin was entirely predictable, if not inevitable. From the state's original strike that resulted in commercial production, which occurred in Corsicana in 1894, Texas had been infected with oil fever. Not only the great Gulf Coast fields—Spindletop, Saratoga, Batson, Sour Lake, Humble, and Goose Creek—of the early years of the century, but also significant later discoveries closer to West Texas, such as Ranger in 1917 and Desdemona and Breckenridge in 1918, carried the fever. Then in 1920, one of the great fields opened in Mexia, thirty miles south of Corsicana.

Even before Corsicana raised the curtain of commercial production, drilling had begun in the Permian Basin. In the late 1880's S. W. Titus drilled an 800-foot dry hole near San Angelo, and C. W. Post put down a well in 1910 near the city bearing his name, but missed the nearby pool. Most of the oil discovered in the Permian Basin prior to 1920 was a by-product of water wells, and the show of oil encouraged wildcatters to continue drilling. In 1900 a well on the W. W. Turney property near Fort Stockton produced small amounts of oil and gas and continued to do so until 1917. Similarly, traces of oil came from wells around Toyah, in Reeves County, and around Carlsbad and Artesia, in Eddy County. Although none of these wells was commercially profitable, they demonstrated the presence of some petroleum in the area—enough to sustain wildcatters' hopes.

Among the regions where oil prospectors tested their hunches was Mitchell County, in the eastern part of the Permian Basin, well removed from other drilling. Leading citizens of Colorado City, eager for their county to share in Texas' oil wealth, invited potential wildcatters to consider their area. Steven Owen, manager of the Underwriters Producing and Refining Company, a New York corporation, investigated the situation and decided to put down a test well on the Texas and Pacific Railway land, some three miles northwest of the hamlet of Westbrook, nine miles west of Colorado City.

Drillers spudded the well, which became known as T. and P.–Abrams No. 1, on February 8, 1920. The bit encountered a show of oil at 450 feet, but drilling continued to

2,130 feet, where signs of a real well occurred. Work proceeded and the local newspaper reported on June 25 that a big well had come in. At 2,345 feet the bit entered an extremely deep pay horizon (the zone from which oil flowed) of 105 feet. Although the well's output could not be determined at that time, since the tight producing formation needed a shot of nitroglycerine to loosen it and promote the flow of oil, news of the potential Golconda raised lease prices in neighboring portions of the county. The nitro shot failed to produce any definite information about the well, since it caused the hole to constrict. In further drilling the bailer was lost and had to be fished out. Finally, after the well reached 2,530 feet, it began to produce on a pump. Far from being the gusher originally anticipated, the Permian Basin discovery well produced less than twenty barrels daily. It had proved, however, that the region could yield petroleum in commercially profitable quantities and served to spur further exploration.

Of the several wells drilled in the area of T. and P.–Abrams No. 1, the one that really documented the worth of the field was Morrison No. 2, also drilled by Underwriters Producing and Refining. Spudded in October, 1920, it hit oil at 2,946 feet in February, 1922. The Rio Grande Oil Company of El Paso bought the production from these wells and built a two-inch pipeline to the railroad at Westbrook. This first pipeline in the Permian Basin began carrying oil at the end of March.

While inaugurating commercial production of oil in the region, Mitchell County never underwent a genuine boom. No bustling town sprang up near the Westbrook field or the other producing areas of the county. The reason was simple: the amount of oil flowing from Mitchell County wells did not encourage a sizeable influx of population. There was no frenzied drilling, no forest of derricks. Even though there were no boom towns, the county seat, Colorado City, became a minor oil center. The California Company, a Standard subsidiary that took over the Underwriters' properties, had its Texas headquarters there, and the Col-Tex Refinery began operations just west of town in 1925. This refinery handled 10,000 barrels a day until the 1950's, when Cosden Petroleum Corporation bought it, transferred operations to its Big Spring plant, and closed the Colorado City refinery. Until that time, the refinery had employed 140 men and had met an annual payroll of around $350,000.

Oil-field children playing on the coiled rope at T. and P.–Abrams No. 1. Billy Thomson, son of the driller, is on the right. *Texas Tech, Southwest Collection.*

Left to right: James M. Charlton, Samuel A. Sloan, W. J. Vaughn, and Steven S. Owen, leaders in developing Mitchell County oil fields. *Texas Tech, Southwest Collection.*

Like the discovery well, Morrison No. 2 was also located a few miles northwest of Westbrook, but it proved to be a much more substantial producer. Standing at the derrick, *left to right,* are Charles Maule, a Texaco scout; Steve Owen, representing the Underwriters Producing and Refining Company, owner of the well; Dan Lewis, driller; and Jerry Ponsia of Reno, Nevada, representing investors in the well. *Abell-Hanger Foundation Collection, PBPM.*

In Mitchell County, Ed Doughit No. 1 formed the nucleus of a community of tar-paper shacks. These 1922 living quarters typified new oil fields. *Abell-Hanger Foundation Collection, PBPM.*

Water for the boilers of T. C. Richardson No. 1 came directly from the Colorado River. This wooden skidway raised and lowered the pump as the water level changed. *Abell-Hanger Foundation Collection, PBPM*.

The first well the California Company (Standard Oil of Texas subsidiary) drilled in the state was T. C. Richardson No. 1 on Morgan Creek in Mitchell County, September, 1922. Assembled on the derrick floor were, *left to right, kneeling*: Len Wallen, tool dresser; Earl Cramer, drilling foreman; Mrs. Moore; Mrs. Earl Cramer; Mrs. Litton; Mr. Litton; and Tom Stoneroad, president, City National Bank, Colorado City; *standing*: Al Giavochini, tool dresser; Reginald Gary, roustabout; T. C. Richardson, on whose land the well was drilled; Harry Smith, driller; Mr. Buchanan, local rancher; Mr. Moore, bookkeeper; Max Thomas, president, Colorado National Bank; Judge C. H. Earnest; Walter Vane, landman; William Simpson, superintendent (holding the Moore baby); Joe Smoot, vice-president, Colorado National Bank; woman with hat unidentified; Mrs. Ledger Smith; Ledger Smith, driller; and Willard Classen, geologist. Despite the auspicious turnout, the drilling produced only a dry hole. *Fannie Bess Sivalls Collection, PBPM*.

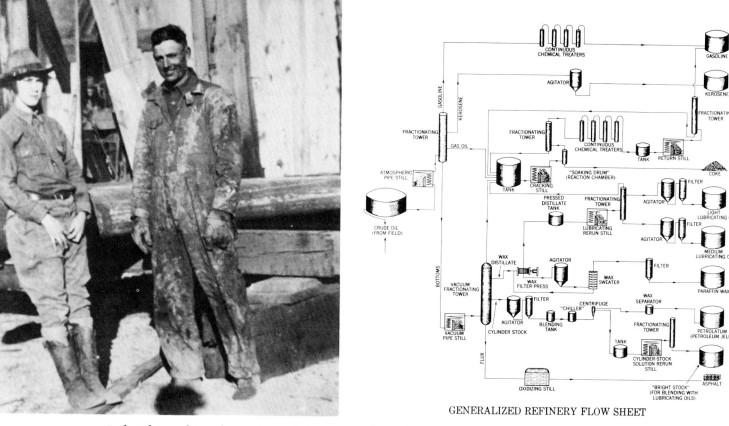

GENERALIZED REFINERY FLOW SHEET

Left: Ed Doughit and an unidentified woman at the rig bearing his name, 1922. *Abell-Hanger Foundation Collection, PBPM. Right*: This 1933 flow chart illustrates the operations of early refineries, such as the Permian Basin's Col-Tex and others. *U.S. Bureau of Mines.*

As Mitchell County proved itself as an oil producer, Colorado City became a refining town with the building of the small Col-Tex Refinery in the 1920's. *Abell-Hanger Foundation Collection, PBPM.*

W. H. Badgett No. 1, near Buford in Mitchell County, gushing thick brine in 1925. In addition to killing the corn field where the derrick stood, the salt water destroyed vegetation in the surrounding forty acres. *Eleanor Wheeler Collection, PBPM.*

The weight of the dried brine caused the collapse of Badgett No. 1's derrick, and the encrustation of salt on the wrecked rig resembled a snowfall. *Eleanor Wheeler Collection, PBPM.*

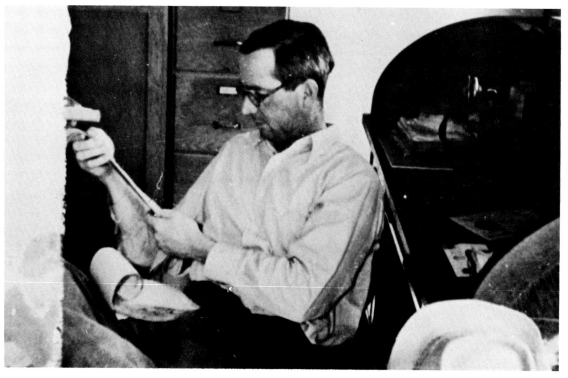

Frank H. Kelley, a landman for the Magnolia Petroleum Company, studying a well log at his office in Colorado City. *Texas Tech, Southwest Collection.*

University Lands and the Santa Rita

As beneficial as oil was to the economy of Mitchell County, the Permian Basin would not be known as one of the world's great petroleum reservoirs had it not been for further, and far more substantial, discoveries. The first of these occurred on University Lands in Reagan County. Since that strike was highly important to the economies of both the Permian Basin and the University of Texas, it is important to understand the origins of the University Lands.

In 1838 Mirabeau B. Lamar, president of the Republic of Texas, urged its congress to appropriate some of the public domain to support higher education. The next year it specified that fifty leagues, or 221,400 acres, be set aside for this purpose, and in 1858 the state legislature upped the figure to one million acres. According to the 1858 law, the acreage included good agricultural land. Since no university had been established by the end of Reconstruction, framers of the 1876 constitution felt no pangs of guilt in shifting University Lands away from the valuable farming region into the arid plains of West Texas. The contiguous counties involved were Schleicher, Crockett, Terrell, Pecos, Upton, Reagan, and Irion. Then in 1883 the legislature appropriated a second million acres, including land in what are now Andrews, Crane, Culberson, Dawson, Ector, El Paso, Gaines, Hudspeth, Loving, Martin, Ward, and Winkler counties. The University Board of Regents appointed O. W. Williams of Fort Stockton to survey the grant. Thus, when the University of Texas opened in 1883, it had an endowment of vast lands of dubious worth.

Since the 1876 constitution prescribed the Agricultural and Mechanical College as a branch of the university, A&M qualified for a share of the income from University Lands. From the 1880's intermittent, sometimes stringent controversies arose between the two institutions over this income. When it became evident that the University of Texas was receiving handsome oil royalties (the lessor's standard one-eighth), A&M's interest intensified. Berte Haigh, the geologist who for many years directed the Office of University Lands, contended (perhaps with a shade of UT bias) that only with the oil royalties did A&M "suddenly remember" its constitutional status and demand 50 percent of the income. In 1930 the boards of the two institutions agreed that A&M would receive one-third of the income from the investment of the Permanent Fund, since it already got federal assistance as the state's land grant college. The state constitution proscribes expenditures from the Permanent Fund, so only its interest income has been available for higher education. And that has been restricted to capital improvements. Nonetheless, the benefits of University Lands to higher education have been immense. In the 1974–1975 academic year, 4,079 oil and 154 gas wells on University Lands provided $66,642,919.61 in royalties. Cumulative income as of April 30, 1981, was $1,408,411,346.92.

Any disappointment the University of Texas may have felt over the transfer of its lands to West Texas did not deter it from making the best of the situation. It leased land for grazing and realized a little additional income from surface easements and the sale of sand, caliche, and gravel. The fact that at the turn of the century oil was revolutionizing the Texas economy prompted the University's Bureau of Economic Geology to investigate the possibilities of finding petroleum on University Lands. For years Dr. Johan A. Udden, chief of the bureau and a professor of geology, studied the region, and in 1916 he informed the Board of Regents of the likelihood of finding oil on University Lands. He accompanied his letter with a map showing where deposits of petroleum might be found, including the southwestern corner of Reagan County. Udden shortly published his study, so the information was available to anyone interested.

Such a person materialized in Rupert P. Ricker, who had arrived in Reagan County as a sixteen-year-old youth in 1906. His family had traveled from Corsicana in a covered wagon. The father's ranching venture failed and he moved to Big Spring, where he worked in the Texas and Pacific Railway shop. After a few years as a brakeman on the line, young Ricker enrolled in the University of Texas and graduated in law in 1915. During World War I he served as captain of an army intelligence unit at Fort Sam Houston. Sergeant Frank Pickrell was a member of his unit. After the war, Ricker determined to seek his fortune in Reagan County. He set out to practice law, but, with his father's dismal experience in ranching as a reminder, Ricker decided not to try to supplement his income by raising cattle. He began to consider the possibilities of oil beneath the barren stretches of Reagan County. To get some reliable information on the subject, he went to his alma mater and read Dr. Udden's reports. Armed with copies of Udden's publications, Ricker returned to Big Lake, opening a law office and dreaming of gushers. With few legal clients, Ricker had plenty of time to devote to his plans. He hoped, with the aid of some associates, to get drilling permits on a wide swath of University Lands.

Ricker and his associates filed 171 applications with county clerks for drilling rights in 674 adjoining sections, or 431,360 acres. To receive drilling permits, Ricker had to file the applications with the General Land Office in Austin and within thirty days pay the fee of ten cents per acre, or $43,136. This figure far exceeded what Ricker and his associates could raise. Consequently, Ricker traveled to Fort Worth in 1919, hoping to interest some oilmen there in his scheme, but to no avail. Quite by chance, on the streets of Fort Worth he encountered Frank Pickrell. Pickrell, from El Paso, had similarly determined to enter the oil business, knowing nothing about it but assuming that it offered a young man the best chance of getting ahead at that time. With Pickrell was Haymon Krupp, a wealthy El Pasoan who also wanted to invest in oil.

Upon learning of their mutual interest, Ricker told Pickrell and Krupp that he had "the hottest deal anybody ever heard of" and invited them to his hotel room to learn the details. There he showed them his documents: maps, applications for drilling permits, and the pertinent state laws. Ricker admitted that he had been frustrated in trying to raise the needed money, so he offered to sell his and his associates' interest in the entire project for $50,000. Krupp and Pickrell agreed to consider the proposal and after a few hours countered with an offer of $2,500, recognizing Ricker's desperation as well as the long shot they were taking. Having little alternative, Ricker accepted the offer, since it enabled him and his group to recoup their expenses. Thus the man whose vision led to the first great oil field in the Permian Basin exited from the scene with no share in the profits.

After Krupp and Pickrell refiled the applications, Krupp began to raise money to finance the exploration for oil. Krupp organized the Texon Oil and Land Company and named Pickrell vice-president and general manager. The New York curb market listed the company's stock, but few buyers were interested, so Pickrell began trying to peddle it himself. He was sometimes able to get forty cents for a one-dollar share, par value. With the delays in financing, the company had to request that the state legislature extend the time by which drilling was supposed to commence. The public land commissioner, J. T. Robison, eager to derive revenue from University Lands and convinced that the company actually wanted to drill and not just speculate in stock, persuaded the legislature to amend the law. The amendments extended the time for drilling and provided for the combination of drilling permits into one sixteen-section unit. This combination enabled the company to sell stock more easily and to secure needed capital.

With $137,000 raised, Texon undertook its first well. Some Roman Catholic women in New York who had invested in the company requested that the well be named after Saint Rita, patron of the impossible, since the entire venture seemed so unlikely. They moreover gave Pickrell some dried rose petals blessed in the saint's name with which to christen the well. He honored their request. After the derrick was finished, he climbed to the top, dropped the petals, and named the well Santa Rita No. 1, making linguistic allowance for Hispanic influence in the region.

Preparations for drilling this well were complicated by the facts that there had been no prior exploration in the area and that the remote location compounded logistical problems. To select a site for the well, Pickrell engaged the services of Hugh H. Tucker, a self-trained geologist, who assured Pickrell that he knew well the geological structure of southwestern Reagan County. Pickrell agreed to pay him $100 per day for five days, a handsome sum for 1919, hoping that Tucker could choose the right place for the well.

Tucker believed the oil-bearing structure was nine miles wide and thirty miles long and recommended drilling three and one-half miles southwest of the Orient railroad siding at Best, the only place from which equipment could be unloaded and hauled to the well site. Pickrell, recognizing the great expense of transporting heavy drilling equipment over the unmarked desert, reasoned that if the structure were as extensive as Tucker thought a more convenient drilling spot would be possible. Tucker disagreed with Pickrell's selection, but the latter decided to follow his own judgment. Ironically, later drilling on Tucker's original site produced a dry hole.

According to the provisions of the drilling permit, the Texon company had to initiate drilling before January 9, 1921, or forfeit the opportunity. Pickrell, so occupied with the sale of stock in New York, let the time slip up on him. Realizing that time was running out, he wired Land Commissioner Robison to ask if beginning a water well would satisfy the letter of the law, since water was vital to the cable-tool process he would use with Santa Rita No. 1. Robison replied affirmatively, and Pickrell bought a decrepit water well drilling machine in San Angelo. He prevailed on the Orient railroad to delay its freight train west for half a day so that he could load all his equipment, including a wagon and horses to pull it. The railroad complied, and the train reached Best in the afternoon of January 8, putting the car carrying Pickrell's equipment on the siding. After loading the drilling machine on the wagon, Pickrell and two helpers headed west four miles to where Pickrell had staked the well. By eight o'clock the well was under way, and Pickrell stopped a startled motorist to get him to sign an affadavit that drilling had begun before midnight. This water well, completed within three months, was eighty-seven feet north of the railroad tracks, and the Santa Rita stake was another eighty-seven feet farther north.

As the water well was drilled, Pickrell hired Carl G. Cromwell for fifteen dollars a day as the driller for Santa Rita. Cromwell had had wide and successful experience as a driller and proved an excellent choice. The two men traveled to Ranger, where they bought drilling equipment at 10 percent of its value, since that field was waning. On August 17, 1921, Santa Rita No. 1 was spudded. During the twenty-one months of drilling, Cromwell, his wife, and little daughter lived in a twelve-by-twenty-four-foot wooden shack nearby. The isolation of the place weighed heavily on the Cromwell family, as well as the drilling crew, which had a steady turnover because of the monotony of the work. Pickrell was constantly pressed to raise funds to pay the crew and buy casing so that drilling could continue.

Through the dreary months the bits pounded away until the well reached 3,050 feet on May 25, 1923. About 5 P.M. Dee Locklin, the tool dresser, noticed gas bubbles at the top of the casing. He and Cromwell pulled the tools and bailed the well several times, excited by the impressive showing of oil. Confident that the well would produce, they boarded up entrance to the derrick floor and scurried around to neighboring ranches, leasing mineral rights to some 30,000 acres. Early on the morning of May 28, with no further human encouragement, Santa Rita No. 1 blew in, sending heads of oil over the crown block. Because the Texon company had not expected the well to flow in such quantity—the daily heads exceeded sixty barrels each—it had not prepared adequate storage facilities. Consequently, until June 25 the well flowed out of control. On June 10 the Orient railroad organized an excursion train from San Angelo to view the well, which dutifully gushed forty-three minutes for the visitors, many of whom doubtless were persuaded to invest in the Texon company. In addition to coming by train, viewers arrived in at least 200 automobiles from various parts of West Texas. Among those who visited the well was Dr. Udden, taking justifiable pride in the soundness of his scientific observations.

Once Santa Rita opened the Big Lake field, its impact quickly became evident in the region's economy. Initially, the field ensured the survival of the Orient railroad, which would serve as the sole transportation link to the field. The village of Big Lake, fourteen miles east of the well, experienced rapid growth, since its 200 citizens offered the only pretense of a town in Reagan County. Big Lake had the institutions around which communities develop: a bank, stores, a hotel, a school, and two churches. Best, the railroad siding just four miles east of the well, developed into something of a supply center and aspired to become a wide-open town—if one that belied its name. As long as boom conditions prevailed in the field, Best survived, but it could not outlast the development phase of the field.

When it became obvious that Santa Rita No. 1 had tapped the treasure of a vast petroleum reservoir, the Texon Oil and Land Company knew that it did not have sufficient resources to develop the entire field and sold a sixteen-section lease right to Michael L. Benedum, "the great wildcatter" from Pittsburgh. Benedum, who had extensive holdings in Pennsylvania, West Virginia, Oklahoma, Illinois, Louisiana, Kansas, Mexico, and the Philippines, had been instrumental in developing the Desdemona field in 1917–1918, so was no stranger to the Texas scene. He formed the Big Lake Oil Company to develop the field, reserving one-fourth of its stock for the Texon company, which appointed two of the seven directors. Benedum's close associate Levi Smith became president and Frank Pickrell, vice-president. To undertake the systematic development of the field, Smith imported drilling crews and their families. He built the company town of Texon near the field, thus adding to the urbanization of that frontier area. Wanting Texon to be a model company town, Smith built 242 neat frame houses with electricity, running water, and natural gas. He planted trees, shrubs, and flowers along the streets and in yards to relieve the natural harshness of the terrain. For recreation Benedum and Smith imported a semi-professional baseball team. Accentuating the family atmosphere of this model town, only 38 of the 249 Texon employees were bachelors. Smith, like Benedum, was religious and taught a Sunday school class in Texon. The community lasted only into the 1930's. Once development was finished, the workers moved away, leaving the town to atrophy. Now a house trailer or two is all that remains of a once bustling community.

Profiting most from the Big Lake oil field was San Angelo, eighty-three miles to the east and north. In 1923 San Angelo had a population of 14,000 and was a trade center for much of West Texas. Situated on the Concho River, it was headquarters for the Texas wool industry. As the nearest town with any office facilities, San Angelo naturally attracted oil companies operating in the Big Lake field, as well as others engaged in further exploration in the Permian Basin. Another reason the town became a significant oil center was that Houston Harte owned and published the San Angelo *Daily Standard*. Intensely interested in the oil business, Harte developed a distinguished oil page, and his journal became the best record of the regional industry. Along with the general prosperity engendered by the Big Lake field, the division headquarters of the Orient railroad in San Angelo expanded as traffic to the field increased enormously. But the town was not destined to be the dominant city of the Permian Basin—it lay too far east of the center of production. Instead, that dominance would come to be shared by two communities, Midland and Odessa, situated in the middle of the region.

The Texas Constitution of 1876 assigned land in Pecos County and elsewhere in the Permian Basin as part of the endowment for the University of Texas. In August, 1901, the University's Mineral Survey photographed this part of its terrain looking toward the courthouse in Fort Stockton. Much of the University Lands looked like this before oil was discovered on them. *UT, Humanities Research Center.*

Left: Johan August Udden, 1859–1932, was a Swedish immigrant who became director of the University of Texas' Bureau of Economic Geology. His studies of the Permian Basin furnished lithologic and stratigraphic data that led to oil exploration on University Lands. *Samuel D. Myres Collection, PBPM.*
Right: In 1916 Udden submitted this map in a report to the University Board of Regents. The circles indicating the possibility of oil demonstrated his scientific acumen, since the first oil struck on University Lands was in Reagan County. *Samuel D. Myres Collection, PBPM.*

University Lands survey crews continued their work in the Permian Basin long after the discovery of oil. In Pecos County in 1932 were, *left to right*, *standing*: Frank F. Friend, Kelly Treadway, Pinkie Pittman, Bill Conklin, and John Oliver; *kneeling*: Norris Creath, Jess Conklin, Kirren Oliver, and Gib Poteet. *Abell-Hanger Foundation Collection, PBPM*.

Left: Big Lake attorney Rupert P. Ricker studied J. A. Udden's reports and filed for drilling permits on 431,360 acres—all of the University Lands in Reagan, Irion, Upton, and Crockett counties—in 1919. *Samuel D. Myres Collection, PBPM*. *Right*: Frank Pickrell had served in Captain Ricker's company in World War I. He and Haymon Krupp bought out Ricker's interest in the project for $2,500. *Abell-Hanger Foundation Collection, PBPM*.

Santa Rita No. 1 blew in May 28, 1923, a harbinger of vast wealth lying beneath the University Lands, not only in southwest Reagan County, but elsewhere in the Permian Basin. *American Petroleum Institute*.

Left: Shortly after Santa Rita No. 1 opened the Texon field, excursion trains brought prospective investors out from San Angelo, and people drove to the well from considerable distances. Such a train is visible in the right background, as are automobiles to the left. *Anton Theis Collection, PBPM. Right*: Haymon Krupp combined with Pickrell, also from El Paso, to form the Texon Oil and Land Company, which undertook to drill Santa Rita No. 1. *Abell-Hanger Foundation Collection, PBPM.*

Driller Carl G. Cromwell's family lived in a small house at the well site during the twenty-one months of drilling. Here they peer into Santa Rita No. 1's slush pit. *Abell-Hanger Foundation Collection, PBPM.*

Oil men and officials of the University of Texas gathered at Santa Rita No. 1 in 1934 to celebrate the wealth accruing from University Lands. The overhead sign dated the well's completion one day too early. *Front row, left to right*: W. M. Griffith, assistant superintendent, Big Lake Oil Company; Jess Conklin, University Lands surveyor; Ed Warren, superintendent, Texon Oil and Land Company; Frank F. Friend, University Lands surveyor; J. S. Posgate, engineer, Big Lake Oil Company; Charles I. Francis, UT regent; K. H. Aynesworth, UT regent; Beauford Jester, UT regent; Hal P. Bybee, geologist in charge, University Lands; and D. R. Johnson, superintendent, Big Lake Oil Company; *back row, left to right*: H. Lutcher Stark, UT regent; V. Stell, gauger, University Lands; Nalle Gregory, geologist, University Lands; Charles E. Beyer, general manager, Big Lake Oil Company; H. Y. Benedict, president, University of Texas; and E. J. Compton, land agent, University Lands. *Anton Theis Collection, PBPM.*

After Santa Rita No. 1's original wooden derrick and rig were dismantled, they were replaced with the above equipment. The steel derrick serves only as a historical monument, since derricks are superfluous once production begins. The well continues to produce oil. In 1958 the original wooden rig, minus the derrick, was installed as an outdoor museum on the campus of the University of Texas at Austin. *Texas Mid-Continent Oil & Gas Association.*

After Santa Rita No. 1 blew in, the Texon Oil and Land Company erected this office in the settlement that took its name from the company. In the mid-1920's, gathered in front of the office were, *left to right*, John O. Carr, G. C. McDermitt, W. W. Pittman, Judge W. B. Moore, E. G. Cauble, Sr., Guy Sowell, Joe Moore, and an unidentified man. *PBPM.*

Lacking funds to develop the Big Lake field, Texon Oil and Land Company sold its lease right in sixteen sections of the field to a new corporation, the Big Lake Oil Company. Heading this new organization was Michael L. Benedum, "the great wildcatter" from Pittsburgh. He and the lieutenant who ran the Big Lake Oil Company, Levi Smith, are pictured along with officials of other companies controlled by Benedum. Those identifiable are, *left to right,* (1) James Chaplin, president of Colonial Trust of Pittsburgh, (4) W. B. Hines, (5) Levi Smith, (6) Ted Williams, (8) Michael L. Benedum, (10) Foster Parriott, president of Transcontinental Oil Company, (11) A. B. Dally, (13) Henry B. Davenport, director of Plymouth Oil Company, (14) Tom Robson, and (15) Fred Miner. To combat the social isolation of Texon, Benedum organized a company baseball team and built a park, where he and his aides gathered on May 23, 1926. *Mrs. Roy Gardner Collection, PBPM.*

The uniformed Oilers team in Denver after winning a national semi-pro tournament in 1928. The Big Lake Oil Company hired Roy Gardner, *fifth from left, top row,* right out of college just to play ball. Later he worked for the company fulltime. *Mrs. Roy Gardner Collection, PBPM.*

At the Big Lake Oil Company's cashinghead gas plant in Texon, 70 percent of the gasoline produced needed no refining, September, 1925. *Institute of Texan Cultures.*

Some thirteen miles east of Texon, Big Lake was the only town of any size in Reagan County. The Blackwell Oil and Gas Company built a camp near Big Lake for its workers, pictured here in 1927. *John J. Kovach Collection, PBPM.*

Left: This Big Lake Oil Company gusher helped make 1925 a banner year for the company. *Institute of Texan Cultures*. *Right*: Originally the Big Lake Oil Company's University No. 3-C at Texon was a shallow well, but when deepened to the Ellenburger formation in 1930 it produced this jet of gas and salt water. *Mrs. Roy Gardner Collection, PBPM*.

By 1925 the Big Lake field had its "golden lane," like that of Mexia. *Institute of Texan Cultures*.

For decades operators in the Big Lake field disposed of salt water from the wells the easiest way, by running it out on the ground. By 1970, this practice had turned the field into a sterile, alkaline desert. *Institute of Texan Cultures.*

Best, four miles east of Texon, aspired to be the urban center of the Big Lake field. Despite this start in 1924, it never rivaled Big Lake. The lone gasoline pump marks the primitive filling station. *PBPM.*

Operating out of Best, Continental Supply Company salesmen E. M. Johnson and John J. Kovach rest on the running board of their roadster, stopped in the Big Lake field in 1926. Note the company name on the spare tire cover. Two spares were common, because mesquite thorns easily punctured the primitive tires of the day. *John J. Kovach Collection, PBPM.*

Main Street in Big Lake, 1932. *PBPM.*

Flynn-Welch-Yates No. 3 struck oil at 1,947 feet on April 9, 1924, establishing the Permian Basin's third commercially successful oil-producing area. This well has often been misidentified as Illinois Producers No. 3, since that company was a forerunner of Flynn-Welch-Yates. *UT, Archives.*

Artesia Field, New Mexico

WITH the opening of the Artesia field in 1924, geologists' calculations that the Permian Basin extended into New Mexico were proven, although this understanding was not immediate. Geological boundaries turned out to be much more cohesive than geographical ones, as the economy of southeastern New Mexico became one with that portion of West Texas in the Permian Basin. When Flynn-Welch-Yates No. 3 began flowing on April 19, 1924, it inaugurated the Basin's third field, located approximately sixteen miles southeast of Artesia. Prior to this discovery Van S. Welch, managing the Illinois Producers Company, had already drilled two dry holes around Dayton, south of Artesia, and was preparing to move his equipment to Burkburnett. By joining him in the venture, Martin Yates, Jr., a realtor who dealt in oil and gas leases around Artesia, persuaded Welch to drill in the area east of the Pecos River. Thomas Flynn, one of the partners in Illinois Producers, did not actively participate in the development of the Artesia field, yet he and Welch owned three-quarters of the new business. Their first two wells produced gas, but insufficient oil to be commercially worthwhile.

Since the company's geologist, V. H. McNutt, had not staked appropriate locations for these wells, Yates wanted his wife, whose intuition he trusted, to choose the site of the third well. She did, and a gusher resulted when the bit hit 1,986 feet. Hoping to increase the production from the first month's 500 barrels, the owners hired Tex Thornton of Amarillo to shoot the well with 135 quarts of nitroglycerine. No spectacular improvement resulted, but the well regularly produced 20 to 50 barrels per day—characteristic of the modest but reliable wells in the Artesia field. At the end of May, 1926, the company had forty-eight wells, each yielding an average of 19.59 barrels daily. Many other operators, such as the Danciger, Ramage, Ohio Oil, and Texas companies participated in developing the field.

A year after the opening of the Artesia field, the test well of the Maljamar field struck oil at 1,989 feet. This well was only a mile northeast of Flynn-Welch-Yates No. 3. Although Maljamar might look like an old Spanish place name, in fact the field took its name from those of the owner's children: Malcolm, Janet, and Marjorie Mitchell. Although the Maljamar corporation's leases were small, in the spring of 1925 its five wells yielded approximately 1,500 barrels per day. More important, the Maljamar production stimulated numerous offset wells on the west and north boundaries of its lease.

When oil was discovered in April, 1924, Artesia was a community of 1,900 people serving the surrounding farming area. The town took its name from the plentiful artesian wells that ensured a prosperous local economy. Because Artesia was the only town in the area, it eagerly responded to the new developments. The Chamber of Commerce and newspaper made clear that they welcomed the growth oil would bring. Thus, while oil did not create a boom town at Artesia, it enabled a stable community to grow and prosper. Evidence of Artesia's welcome of the new industry came with the efforts of businessmen to raise money to build a refinery in town. In an indirect way, petroleum exploration accounted for further town building in Eddy County. V. H. McNutt, while drilling for oil some fifteen miles southeast of Carlsbad, struck a potash deposit at 600 feet. The potash bed proved to be 1,900 feet in depth, making it the largest found in the United States. This discovery resulted in Carlsbad's becoming the center of the potash industry in the country.

Drilling New Mexico's discovery well early in 1924 are, *left to right*: James W. Berry, with the big hat; Basil Tigner in front of Berry; Ed Wingfield; and Frank Tigner. Located about fifteen miles southeast of Artesia, Flynn-Welch-Yates No. 3 bore the name of the operating company, since the land was not privately owned. *Abell-Hanger Foundation Collection, PBPM*.

Left: On June 29, 1924, Tex Thornton of Amarillo shot Flynn-Welch-Yates No. 3 with 135 quarts of nitroglycerine, resulting in an initial production of a thousand barrels, which dwindled to a steady flow of twenty to fifty barrels daily. Here Thornton holds a torpedo before lowering it into the well. This photograph came from a different well, but the process was the same. After shooting Flynn-Welch-Yates No. 3, Thornton performed one of his legendary feats. A head of oil and gas hoisted two torpedoes out of the hole, but the alert Thornton grabbed each, saving his own hide, plus those of the drillers, and the well itself. *UT, Archives. Right:* In the Artesia field, Flynn-Welch-Yates drilled Jackson NO. 2 for the Windfohr Oil Company with a patented turnbuckle derrick, which had steel rods reinforcing the wood. Turnbuckles tightened the rods to brace the derrick. Such derricks phased out in the early 1930's. *Abell-Hanger Foundation Collection, PBPM*.

A blowout of Robinson No. B-1 on the American Republic Corporation lease in the Artesia field, 1930. *Blowout* is a general term referring to any uncontrolled production of oil, gas, or water. *Frank Forsyth Collection, PBPM.*

Van S. Welch and Martin Yates, Jr., in Welch's Artesia office, September, 1948. Their obvious good spirits could have resulted from their third decade of success in the petroleum business. *Abell-Hanger Foundation Collection, PBPM.*

The Navajo refinery, Artesia, August, 1975. *Samuel D. Myres Collection, PBPM.*

Left: The Hopi Drilling Company's cable-tool rig drilling Orla-Petco Ferguson No. 1 a few miles east of Carlsbad, July, 1980. *Richard Donnelly. Right:* Near Orla-Petco Ferguson No. 1 in Eddy County is the Sharp Drilling Company's giant rotary rig for the Coquina Oil Corporation's Carlsbad-Pecos No. 1, offering a striking contrast to Hopi's smaller, less complicated cable-tool outfit. Pipe is standing in the rack in thribbles, that is, three joints screwed together. Although drillers add one joint of thirty or more feet at a time, when they pull the drill string out of hole to change bits, they break it into thribbles for convenience in racking and rerunning. July, 1980. *Richard Donnelly.*

Roy Hill, driller on Orla-Petco Ferguson No. 1, July, 1980. In the center a bit stands erect in front of the forge, and the anvil is to the lower left. The big cylinder suspended to the left of the bit is the bailer. *Richard Donnelly.*

An aerial view of the well site of Carlsbad-Pecos No. 1, showing the extensive area required for a modern deep well. In July, 1980, this well had exceeded 12,000 feet in its search for gas. The trailer to the lower right of the rig housed the geologist and his elaborate equipment, while the trailer in the left corner of the area housed the Sharp Drilling Company's tool pusher, head of the drilling crew, and the Coquina Oil Corporation's well superintendent, the "company man." Such men are on the well site twenty-four hours a day. *Richard Donnelly.*

Daily drilling report for Carlsbad-Pecos No. 1, July 17, 1980. The upper left corner shows the signatures of A. L. Crain, the company man, and Dwayne King, the tool pusher. On the far right, each driller has signed for his tour, or shift. From the beginning, those in the oil field have pronounced the word as if it were spelled "tower." *Richard Donnelly.*

Artesia Field / 41

The World Oil Company, headed by Chester R. Bunker, opened the Permian Basin's fourth field in 1925. This Crockett County field included World No. 1 Cerf Bunker Wildcat No. 4. The temper screw at the base of the threaded column enabled the driller to adjust tension on the drilling line to maximize the effect of the cable tool bit. *Abell-Hanger Foundation Collection, PBPM.*

1925: Strikes in Crockett, Loving, Upton, and Howard Counties

THE year 1925 saw the beginning of oil production in Crockett, Loving, and Upton counties. The Crockett County discovery came when World Powell No. 1 struck oil at 2,625 feet on May 30, 1925. This well lay seventeen miles south of the Big Lake field. In Loving County, Texas' least populated county, the Pecos Valley Petroleum Company's Wheat No. 1 hit pay at 4,212 feet on September 1, 1925. Whereas the Crockett County discovery differed from the norm by being on a sheep ranch, which was owned by L. P. Powell, the Loving discovery came on the more traditional terrain of a cattle ranch, one owned by J. J. Wheat. While both the Powell and Wheat fields were economically profitable, neither was of sufficient magnitude to engender any significant urban development. Aside from some company housing, such as the Lockhart camp in the Wheat field, oil workers in those counties arranged for their own quarters. Driller Ford Chapman, while working in Crockett County, occasionally slept on a derrick floor, although that was a purely temporary arrangement.

Upton County was a different matter, for Johnson and McCamey Baker No. 1, completed on September 25, 1925, at 2,193 feet, prompted the creation of a genuine boom town. Not only did McCamey grow as the field prospered, but Rankin, the county seat, also enjoyed some of the benefits. George B. McCamey, partner with J. P. Johnson in a Fort Worth drilling firm, came to be a major figure in the Permian Basin through a chance meeting with Arthur Stilwell, promoter of the Orient railroad. They found themselves on a train headed for the Big Lake field in 1925, and Stilwell, who had previously purchased considerable acreage in Upton County, urged McCamey to drill. Upon investigating, McCamey found that the Marland Oil Company had already leased over a dozen sections where geologists indicated the presence of oil. When McCamey proposed to drill a test well in exchange for a portion of the holdings, Marland consented, and spudding began on August 20. After the well came in the next month, it produced 192 barrels daily on the pump. In October Johnson and McCamey sold their interest in the well, plus leases on 1,050 acres nearby, to the Republic Production Company for half a million dollars.

McCamey, the town, had in no way a preordained existence, although it was the first entry in the urban sweepstakes near the new field. It began when George Mc-Camey persuaded the Orient railroad to build a switch and siding at the point nearest the discovery well, three miles to the northwest. McCamey agreed to bear the labor expense of construction. When the switch and siding were completed, a railroad employee painted a sign, "McCamey," and posted it on a boxcar; thus did the town get its name. Before the town could really take root, the rival oil-field town of Crossett appeared to outstrip McCamey. But the latter was closer to the wells, and supply firms preferred to locate near the oil field; hence McCamey eclipsed Crossett. Instrumental in promoting McCamey were Woodrow Wilson's postmaster general, Albert S. Burleson, and C. D. Johns, Burleson's brother-in-law.

Boom conditions were at their height in 1927 and 1928, when McCamey reached a population of ten to twelve thousand. During those years, roustabouts earned $5.25 a day and carpenters, $10.00. They paid $35.00 monthly for two- and three-bedroom "shotgun" houses (one room behind the other). Many workers lived in floored tents, and they, like the residents of the houses, used outdoor privies. The town's drinking water had to be shipped in from Big Lake or Alpine in railroad tank cars. Even livestock shunned the alkaline water in the nearby Pecos River, O. W. Williams wrote his son in the summer of 1926. With little rainfall, yellow dust encased McCamey as heavy trucks plowed their way along unpaved streets en route to the field. As in other boom towns, McCamey attracted its share of bootleggers, prostitutes, and gamblers, although their activities were less strident than in other Permian Basin boom towns. With the establishment of churches and schools, McCamey began to put down the roots of permanence. Significantly, there were enough blacks in McCamey to establish an African Methodist Episcopal congregation in 1925. Although the black population of West Texas was negligible before the discovery of oil, blacks, like whites, migrated toward economic opportunity. But the number of blacks in the Permian Basin boom towns never approached the percentage of blacks in the Texas population as a whole. Adding to the permanence of McCamey, Humble established a large camp nearby in 1927, as well as a small refinery. It also moved its regional headquarters there from San Angelo. Before the Depression caused the refinery to close in 1932, it employed 300 men. Revived production in the late 1930's enabled the town to survive.

Somewhat like Colorado City, Big Spring was transformed by oil from a sleepy county seat into something of a city. It served as the commercial center for exploration in Howard and Glasscock counties. The area's first profitable well was Hyer-Clay No. 1, which began producing between nine and eighteen barrels a day on November 13, 1925. More important than the discoveries in the area was the establishment of the Cosden Refinery in Big Spring in August, 1929. J. S. Cosden built the refinery to process crude from his leases on Dora Roberts' land. It became the largest plant in the Permian Basin and now is heavily involved in the manufacture of petrochemicals. The growth of Big Spring offers further evidence that the development of the petroleum industry in the Permian Basin spurred the process of urban development.

Waste oil burning in Crockett County's World pool, 1928. Although this method of disposition was inexpensive and easy, it showed little concern for the environment. *Frank Forsyth Collection, PBPM.*

By March, 1939, this well in Crockett County hit 9,000 feet, reaching into the Ordovician formation. The Ozona-Barnhart Trap Company Wildcat No. 1 was located near Ozona. *Graybeal Scrapbook, PBPM.*

Left: After the Pecos Valley Petroleum Company's Wheat No. 1 struck oil on September 1, 1925, at 4,212 feet, Loving County became the fifth producing area in the Permian Basin. J. J. Wheat owned considerable land in the area of Mentone, which he tried to sell as town lots. He stands in front of his real estate office in 1931. *Abell-Hanger Foundation Collection, PBPM. Right*: Dean No. 1, near Mentone, gushing in 1930. The photographer stood on top of a nearby storage tank. *Ford Chapman.*

In 1926 the Pecos Valley Petroleum Company sold half-interest in Wheat No. 1, plus other leases, to the Lockhart brothers, owners of El Paso's Rio Grande Refinery. They developed the Wheat field and established this H. L. Lockhart camp near Mentone. The Rio Grande Oil Company subsequently became a part of Atlantic Richfield. *Abell-Hanger Foundation Collection, PBPM.*

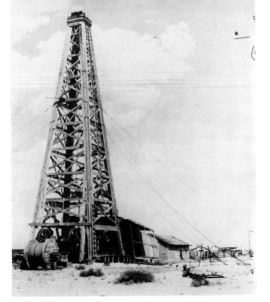

Left: In 1930 the Miller, Burge, Healy, N. F. Chapman No. 1 struck oil on the Chapman-McIlvain lease near Mentone. Robert O. Moorhead later bought the well and in 1966 donated the rig to Kermit, where it now stands in the Winkler County park. *UTPB, Permian Historical Society.* *Right*: George B. McCamey at the discovery well in the field that bears his name, located in southwestern Upton County. *Abell-Hanger Foundation Collection, PBPM.*

From a depth of 2,193 feet, the McCamey well produced 192 barrels per day after its completion on September 25, 1925. As was common with cable-tool rigs, the walking beam used to raise and lower the drilling line was converted into part of the pumping unit. As the motor caused the pitman (the upright beam on the left) to move up and down, the walking beam, rocking on top of the samson post, activated the sucker rod on the right. Its vertical movement pumped oil from the ground. *Shell Oil Company.*

McCamey gave his name to a bustling town, as well as the field. This photograph, dating from the early 1930's, looks east on Fifth Street. *Abell-Hanger Foundation Collection, PBPM.*

Rankin, the Upton County seat, boomed along with McCamey. The corner filling station was a prominent feature on Main Street in 1930. This picture was sent as a postcard, whose message read, "This is a great photo of our wonderful city." *Samuel D. Myres Collection, PBPM.*

Development continued in Upton County in the 1930's. Humble employee Maurice Minnette stands in front of the boiler for J. B. Tubb No. B-1, in the Sand Hills field, April 4, 1938. *Graybeal Scrapbook, PBPM.*

This derrick near McCamey stands between two objects of natural beauty in Upton County, a yucca plant and a mesa, September, 1945. *Texas Tech, Southwest Collection.*

The discovery well in Upton County's Wilshire field. Located on the McElroy lease some twenty-three miles northeast of McCamey, the well hit oil in the Ellenburger formation. This type of "shouldered" rotary derrick was used only briefly during the late 1940's and early 1950's. *Frank W. Lake Collection, PBPM.*

Mr. and Mrs. S. E. J. Cox in Big Spring, celebrating his first well in the Permian Basin, General Oil Company's McDowell No. 1, in 1920. Reflecting the couple's enthusiasm for aviation, she had just landed in a company plane and obviously liked the pose of aviatrix. *Abell-Hanger Foundation Collection, PBPM.*

A 100-quart shot of nitroglycerine caused McDowell No. 1 to erupt on March 26, 1921, but production dwindled to eight barrels a day. The well was about twenty miles south of Big Spring, in northern Glasscock County. Cox, a plunger, had leased 200,000 acres in West Texas, with meager results. To cover his debts, he raised money through mail fraud, for which he received an eight-year prison sentence and a fine of $8,000. *Lee Jones, Jr., Collection, PBPM.*

For Howard and northern Glasscock counties, Big Spring served as the commercial supply center. In the early 1920's W. F. Cushing, a Glasscock County rancher, founded the Cushing Development Company. This scene shows tractors hauling rig timbers from Big Spring to the Cushing ranch for a wildcat well. Lest anyone mistake the purpose of the procession, three signs announced that timbers for the Cushing well came from the Burton-Lingo lumber company. In the background, the Tourist's Garage doubled as a Texaco filling station. *Heritage Museum, Big Spring.*

Howard County's discovery well, Hyer-Clay No. 1, began producing November 13, 1925, yielding between nine and eighteen barrels daily from 1,563 feet. The horses to the right of the well were scooping out a storage pit. *Abell-Hanger Foundation Collection, PBPM.*

The Cosden Refinery began operations in Big Spring in August, 1929. Joshua S. Cosden built it to handle crude from his leases on Dora Roberts' land. The largest refinery in the Permian Basin, it is the only survivor of four that opened at Big Spring. *Mahan & Associates, Inc.*

Left: In the 1950's, Cosden, like many other refineries, developed extensive petrochemical units. An adjunct to the refinery is the UOP Rexforming unit, where pipes for feed stock and products combine with instrument air tubing, electric conduits, and utility lines to create an intricate maze. *Texas Mid-Continent Oil & Gas Association. Right*: In 1961 Fred Hyer, independent oil operator from Fort Worth, returned to his discovery well in the southeastern portion of Howard County. *Abell-Hanger Foundation Collection, PBPM.*

In addition to oil refineries developing petrochemical units, other corporations built plants to manufacture petrochemical products. Next to Big Spring's Cosden Refinery, the W. R. Grace nitrogen
plant gets its feed stock for fertilizer from Cosden. *Mahan & Associates, Inc.*

Crane County

THE banner year for exploration in the Permian Basin was 1926, when wildcatters discovered the most prolific fields of the region: Church and Fields and McElroy in Crane County, Hendrick in Winkler County, Maljamar in Lea County, and Yates in Pecos County. These fields, along with the towns they spawned, have left an indelible mark on the region's history.

The discovery of oil in Crane County is a fascinating story of opportunities, some taken and others missed. George M. Church, heir to the Liggett and Meyers tobacco fortune, wished to speculate in West Texas and teamed up with lease broker Robert Fields. They hired Burton F. Weekley, a drilling contractor from Wichita Falls who had worked for Church previously, to put down a 3,000-foot test well on their lease in Crane County. Never having drilled in the remote sand dunes before, Weekley was reluctant to name specific terms. But when pressed, he replied that he would charge $5 per foot, in addition to an 80-acre offset plot and another 240 acres within the lease. Church agreed, and Weekley spudded Church and Fields University No. 1 on December 28, 1925. Short on cash, Weekley settled his $2,500 bill from the derrick builder by trading him 160 acres—half his lease. This acreage later sold for $100,000. When Weekley owed his driller more than $500, he tendered 80 acres near the well, but the driller turned it down. These acres later brought Weekley $80,000. In mid-February, 1926, at around 3,000 feet, the well first showed signs of oil, accompanied by strong fumes of hydrogen sulfide. Even before the well's completion on April 19, Church and Fields had sold their interest in it, plus other properties in the area, to the Magnolia Petroleum Company for $450,000.

When it became evident that Crane County's discovery well presaged further lucrative strikes, promoters began to build the only town in the county, also named Crane. Beginning at the end of 1926 with simple wooden buildings and hundreds of tents, Crane grew apace as the McElroy field joined with the Church and Fields pool, five miles to the north, to make the county a significant producer. Gulf McElroy No. 1, completed in July, 1926, ushered in a rich new field on land belonging to pioneer rancher John T. McElroy. From around 2,750 feet, the well initially yielded 600 barrels per day, but quickly decreased to 255. Further drilling in the Gulf lease, and other wells, such as Collett No. 4 in 1927, established that the two great Crane County fields were actually one. On the boundary of the McElroy ranch lay the vacancy (an unowned area lying between property lines established by previous surveys) that Edward Landreth leased and drilled with great success in 1927. On these 182.8 acres, shaped like a shoestring, Landreth had sixteen wells of such stellar quality that Texaco purchased his interest in 1928 for $6.5 million.

With the quick growth of the town in early 1927, it became apparent that Ector County should no longer administer Crane County. Upon petition of 87 residents of Crane, Ector County's commissioners court approved the separation, and Crane County became self-governing on September 3. By the end of November, Crane numbered some 2,000 people. As in some other boom towns, such as Borger, most of the businesses were strung out along one street, and Crane's Main Street measured one mile within three months of the town's inception. Company camps near the fields provided better housing than did Crane, a situation that also prevailed in other boom towns. Although Crane had an ample share of rowdies and lawbreakers in its earliest days, it steadily developed the institutions of permanence, such as churches and schools and especially the apparatus of county government. In the process of town building and maturation, county seats have had a natural advantage, which has been obvious in the continuing prosperity of these towns.

In April, 1926, Church and Fields University No. 1 opened production in Crane County. During the spring and summer, a few shacks arose near the well in east-central Crane County, but this photograph, taken on January 7, 1927, indicates the town got started officially December 24, 1926. Ed Baldwin, a Crane Townsite Company official promoting the town, stands on the far left. From the three wooden buildings and five hundred tents then extant, Crane grew to prove the photographer's boosterism valid: It was a town to keep an eye on. *UTPB, Permian Historical Society.*

The Landreth Strip in northeastern Crane County resulted from a vacancy, an unowned area lying between property lines established by previous surveys. J. E. Anderson of Austin and Jax M. Cowden of Midland bought the 182.8 acres from the state in February, 1927, and a month later leased them to the Landreth Production Corporation of Breckenridge. Edward A. Landreth drilled sixteen wells in the strip, which measured 434 feet by 20,000 feet. The wells' prodigious output filled forty-one 55,000-barrel tanks. On June 15, 1928, Landreth sold his interest in the strip to Texaco for $6,500,000. As soon as Landreth proved the strip, the Gulf, Atlantic-Simms, and Tidal oil companies drilled offset wells. An offset well is drilled in an adjoining leasehold as close as possible to the original producer, the lessee hoping to capture his share of the pool. *Jack Nolan Collection, PBPM.*

Left: Burton F. Weekley, driller of the Crane County discovery well, in 1927, taking a core on Collett No. 4, lying between the Church and Fields pool and the McElroy field to the south. Drillers take cores from the bottom of wells to determine whether the rock formation shows prospect of oil. Cores are cylindrical columns of rock, ranging from two to four inches in diameter and one to two feet in length. Gulf Oil and others operating in Crane County tried to prove with Collett No. 4 that the Church and Fields field and the McElroy field were all one pool. *Abell-Hanger Foundation Collection, PBPM. Right*: An Eastland Oil Company well in the McElroy field, 1927. *Richard Donnelly.*

After Landreth sold out, Texaco put its star on the wells, as at Cowden-Anderson No. 14, which in 1929 yielded 375 barrels a day. *Texaco Archives.*

Gulf drilled the discovery well on the McElroy ranch on July 21, 1926, and developed the field over the next several years. The drillers of Gulf McElroy No. 103, which began March 21, 1933, experimented with Hughes Tool Company's new "tri-cone" bit. During the twenty-six months of drilling, the well chewed up 965 bits as it hit 12,786 feet—a record for the time. The Hughes company learned valuable lessons that enabled it to develop modern rock bits. *Abell-Hanger Foundation Collection, PBPM.*

The Gulf camp at Crane, January 25, 1935. No. 103 is the tall derrick in the center background. *Abell-Hanger Foundation Collection, PBPM.*

Shortly before Gulf McElroy No. 103 began producing on May 25, 1935, its drilling crews assembled on the derrick floor. Atop the can of Gulf Supreme cup grease sat a Hughes tri-cone roller bit. On the extreme left is E. D. Brockett, who later became chairman of the Gulf board. *Abell-Hanger Foundation Collection, PBPM.*

Downtown Crane, the late 1930's. *UT, Archives.*

R. H. Henderson's well in Crane County's Dune field blows out on March 1, 1939. *Abell-Hanger Foundation Collection, PBPM.*

This pumping derrick, built after the well was completed in the Sand Hills field near Crane, is used to pull rods and change pumps, September, 1945. *Texas Tech, Southwest Collection.*

A jackknife rig drilling in the Sand Hills field in the mid-1950's. The deep sand complicated the logistics of road building and hauling heavy equipment. *Abell-Hanger Foundation Collection, PBPM.*

Winkler County:
The Hendrick Field and Wink

THE story of the next big strike in 1926—the Hendrick field in Winkler County—involved more than the search for oil. It involved the manipulations of speculators, more eager to make a fast buck than to find oil. Because the terrain of Winkler County, which forms a ninety-degree angle with the border of New Mexico, was so barren, it proved the ideal locale for speculation. In the 1920's drought had driven Winkler County ranchers to desperation, and they were fair game for any money-raising schemes, including a number proposed by Fort Worth speculators. Knowing the ranchers' immediate need for cash, speculators offered to lease mineral rights on the land for as little as ten cents per acre. After they had developed huge leases, they subdivided them and sold lease rights to five-acre tracts for five to ten dollars per acre.

These speculators also trafficked in potential royalties, purchasing the ranchers' one-eighth for between one and three dollars per acre and then selling that potential royalty for up to fifty dollars an acre. In peddling their schemes—largely through the mails—speculators sometimes offered buyers such slivers as $\frac{1}{128}$s of the original $\frac{1}{8}$ royalty. But if enough suckers paid ten dollars for such a dream, the speculator could make a tidy profit. Somewhere along the line, though, the speculator had to show evidence of developing the property, lest he run afoul of federal mail fraud statutes. One such broker in Winkler County was J. W. Grant, who leased Thomas G. Hendrick's fifty-three-section ranch (33,920 acres) for a dime an acre. Two months later he sold these leases to Roy A. Westbrook for thirty-five cents an acre.

Westbrook, a Fort Worth printer, found oil speculation much more stimulating than his trade. He divided some of his leasehold into 5-acre plots and promptly began to promote them through the mails, saying that he would find oil in the area. Only the most reckless promoter would venture such a statement, since the surface of Hendrick's ranch offered no indication of oil beneath and since the nearest production was fifty miles away. But to fulfill the terms of his mail advertising, Westbrook assigned 2,000 acres, near the 5-acre tracts he had sold through the mails, to the firm of Johnson and Mc-

Camey with the understanding that they must find oil. This firm, which had drilled Upton County's discovery well, then arranged for the Donnellys' Eastland Oil Company to put down the test well. Eastland's payment was lease rights on 400 acres near the test. Eastland spudded the well on February 2, 1926, and on July 17 the 2,525-foot hole began filling with oil. Westbrook's fantastic gamble had paid off—much to his relief, since postal inspectors were breathing down his neck. The well, when deepened to 3,000 feet, began flowing 400 barrels daily.

T. G. Hendrick No. 1 opened an exceedingly rich field, nine miles long, that jolted the sleepy county seat of Kermit, with fewer than 100 people, into lively activity and created the new community of Wink at the southwest corner of the field. Many corporations, such as Marland, Humble, Gulf, Pure, Republic, Atlantic, Magnolia, Texon, Amerada, and Roxana (a Shell affiliate), bought into the field and undertook drilling. Operating under the common-law rule of capture, which held that migratory underground minerals belonged to him who found and "captured" them beneath his property, these companies drilled intensively to get their share of the oil pool. This close drilling of productive wells resulted not only in the frenzied activity that made Wink the quintessential "wild and woolly" boom town, but also in the depletion of gas pressure that lifted oil to the surface and in the long-range diminution of the field's potential.

The Hendrick field marked a turning point in drilling techniques in the Permian Basin, for it was the first field developed largely with rotary equipment. The heavy gas pressures encountered throughout the field militated against cable-tool drilling, since the gas often blew equipment out the hole. Rotary drilling, which controls such pressure with mud pumped into the hole, proved much more practical in the Hendrick field, and subsequent fields were developed largely with rotary equipment, which could make hole faster and deeper than cable tools. Because rotary drilling involved much more complicated equipment, it was accordingly more expensive, running between $45,000 and $65,000 per well.

Contract drillers in the Hendrick field ordinarily worked on two twelve-hour tours or shifts, seven days a week. Those with experience earned from fourteen dollars to twenty dollars a day, and roughnecks got six to eight dollars. Since these workers were unorganized, their job tenure depended strictly upon performance. If someone failed to do his work, he could be promptly discharged. But since reliable workers were always in demand, contractors carried their crews from one well to another in the region. They particularly valued trustworthy and able workers, since climatic conditions and the remoteness of many Permian Basin fields dissuaded all but the hardy—a phenomenon typical of any frontier's process of natural selection.

Those who lived in Wink had to possess some special hardihood, because the town strained to be the most lawless in the region. From its beginning in March, 1927, with only a handful of people, it swelled in 1928 to around 10,000. Included in this number were many honest, industrious citizens, but there were also enough gamblers, whores, thieves, and bootleggers to paralyze the ordinary functioning of a community. Local officials operated in cahoots with the lawless element, so citizens had little redress to the normal institutions of an orderly society. A crusading editor undertook to clean up the town but found she was leading a one-woman fight. Her advertisers deserted her, and no citizens came forward to enlist in the fray. When her hotel locked her out, she realized the battle was lost and left town. After a town commissioner, who ran a bootleg business, murdered a rival bootlegger in 1932 with the assistance of a deputy constable, the county sheriff's office finally cracked down on lawbreakers in Wink. After the court convicted the two felons of murder, lawlessness subsided in Wink, (as has some of the surrounding countryside!). Wink exemplified many of the problems of rapid growth in an area lacking the normal institutions of society. Those stabilizing institutions simply did not develop quickly enough to enable Wink to mature in a constructive fashion.

Even before the discovery of oil on his Winkler County ranch made Thomas G. Hendrick, 1862–1940, wealthy, he was a prominent citizen of West Texas. As Ector County judge in October, 1925, he sat at his desk in Odessa. A Fort Worth newspaper protected the floor from any who might miss the spittoon. During his term as county judge, 1925–1927, Hendrick received $100.00 per month, plus $16.66 as county road superintendent. *Samuel D. Myres Collection, PBPM.*

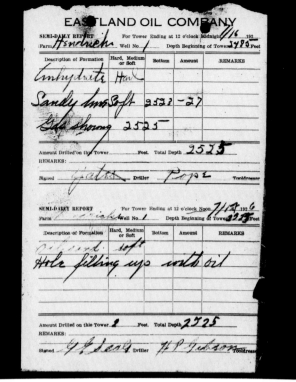

Left: Drilling contractor for T. G. Hendrick No. 1, the Eastland Oil Company spudded the discovery well on February 2, 1926. As the company's semi-daily report for July 16 and 17, 1926, shows, the hole of 2,525 feet was "filling up with oil" on the morning of July 17. Note the phonetic spelling of *tour* on the printed form, and compare this simple document with the elaborately detailed drilling report for Carlsbad-Pecos No. 1, page 41. *Eastland Oil Company Collection, PBPM*. *Right*: Fort Worth oilman Roy A. Westbrook developed the Winkler County discovery well, whose gushing portended a great new field in the Permian Basin. Original production was only 30 barrels daily, but when drilled to 3,049 feet on March 28, 1927, Hendrick No. 1 yielded 400 barrels per day. *UTPB, Permian Historical Society*.

As the Hendrick field developed, automobiles created their own roads through the sand. The driver of the Ward Commercial Photo Company car, bearing a 1927 Oklahoma license, documented that he, too, was at work in the field. *American Petroleum Institute*.

The Wink Townsite Company appealed to the poetic instinct of prospective property buyers. What the company did not list among Wink's possessions was a record of violence reminiscent of Ranger and Desdemona. *Samuel D. Myres Collection, PBPM.*

Town building was one of the most vigorous aspects of the frontier West, and the nascent oil towns of the Permian Basin all looked pretty much alike. In March, 1927, Wink began developing on the southwestern rim of the Hendrick field, which measured more than nine miles on a southwest-to-northeast axis. Located in Winkler County, Wink took its name from the county. *Abell-Hanger Foundation Collection, PBPM.*

From its modest beginnings in March, 1927, Wink mushroomed into a town of some 10,000 by the end of 1928. The congested main street evinces some of Wink's dynamism at the period of its peak population. The Pyote-Wink bus in the foreground linked the new town with the nearest Texas and Pacific railroad station, fifteen miles to the south. While the office on the far left offered townsites, a store across the street sold tents for those who needed immediate shelter. Like most boom towns, Wink suffered a sharp loss of population once drilling was completed. *UTPB, Permian Historical Society.*

Plaza M. Woods's teamster crew in Wink. On the far right, holding the reins, is Harvey Henson. The man with the hat in the Model T on the right is Paul M. Woods, son of the boss. *UTPB, Permian Historical Society.*

Housing for workers in the Hendrick field was so short that teams hauled in prefabricated structures. *Samuel D. Myres Collection, PBPM.*

Companies such as Southern Crude Purchasing, Gulf, and Independent Oil and Gas established camps for their workers right in the Hendrick field. Such camps offered not only better housing than that otherwise available in Wink but also convenience to work. *Frank Forsyth Collection, PBPM.*

The Hendrick field blanketed with snow during the 1928–1929 winter. *Frank Forsyth Collection, PBPM.*

The well in the background was Lion Oil and Refining Company's Hendrick No. 11, a producer. Another company drilled an offset well, in the foreground, trying to tap the same pool, but the derrick collapsed, or "pulled in," while tubing was being hoisted from the hole. *Frank Forsyth Collection, PBPM.*

Left: In the foreground of this tank battery in the Hendrick field are electric treaters for removing basic sediment and water before crude oil was piped into the tanks. *Frank Forsyth Collection, PBPM.* *Right*: T. G. Hendrick No. 2 produced 9,300 barrels of fluid and 500,000 cubic feet of gas daily in March, 1938. At the well site, Humble's eight-by-twenty-two-foot separator segregated oil, gas, and water, piping the last into a pit. Separators are pressure vessels that function principally on gravity but may operate chemically or with heat. *Graybeal Scrapbook, PBPM.*

Left: The Hendrick field extended north toward Kermit, the Winkler County seat. In Humble's Kermit field, N. R. Colby No. C-15 erupts after a 725-quart nitro shot on July 7, 1937. *Graybeal Scrapbook, PBPM. Right*: Colby No. C-15 one minute later. *Graybeal Scrapbook, PBPM.*

On August 12, 1937, the Kermit field's N. R. Colby No. C-16 was ignited by a spark from an electric hand lamp in the cellar. Firefighters are wetting the ground near the well before trying to snuff out the blaze. *Graybeal Scrapbook, PBPM.*

The Keystone field, northeast of the Hendrick field, in Winkler County, November, 1945. The Keystone field proved an important part of Sid Richardson's vast fortune. Born in Athens, Texas, like Clint Murchison, another leading independent oilman, Richardson had extensive holdings in the Permian Basin. *Texas Tech, Southwest Collection.*

Sid Richardson (*center*) with his attorney John B. Connally (*left*) and friend Lyndon B. Johnson. This picture dates from Johnson's term as Senate majority leader in the 1950's. *PBPM.*

The Hendrick pool—above and below ground. For years well operators found it cheaper and easier to dump water onto the ground than to pump it back into the formation, where it could have helped maintain water drive to bring the crude to the surface. On some occasions the only access to wells was by rowboat. *Frank Forsyth Collection, PBPM.*

Some companies, such as Gulf, understood the long-range economies of conservation and built water-injection plants, such as this one at Kermit, to return the water underground. This not only protected the environment but also aided further production through maintaining underground pressure. *Texas Mid-Continent Oil & Gas Association.*

Water injected into the earth may have caused a salt dome to dissolve, creating the Wink sink, pictured here on June 15, 1980. Or this sink could have resulted from two other causes: indiscriminate water flooding of the surface and decades of pumping fluids from beneath the surface. Although the Wink sink garnered great publicity, such occurrences are not rare in oil fields. As the Wink sink developed to the dimensions of 150 by 25 yards, by the end of July, two producing wells nearby had to be plugged and a pipeline rerouted. Part of the original pipeline dangles from the right edge of the sink. *Odessa* American.

When Permian Basin oil producers protected the environment, they enabled the region's original economy, ranching, to continue, as in this scene near Wink in 1950. Many ranchers were initially antipathetic to oil exploration because it often ruined grazing land. *Texas Mid-Continent Oil & Gas Association.*

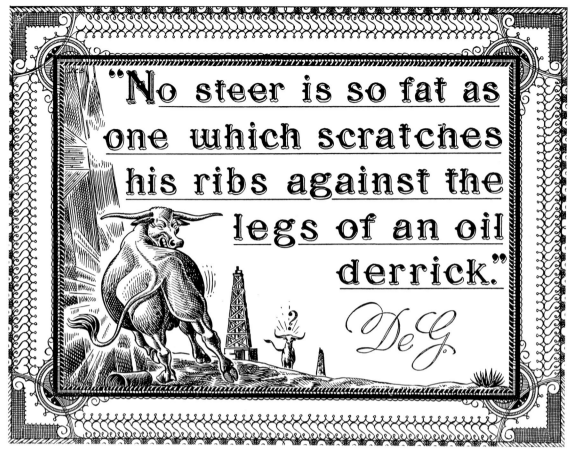

Everette DeGolyer's 1947 Christmas card poked fun at ranchers who scorned the intrusions of oilmen. José Cisneros, an El Paso artist, drew this cartoon for DeGolyer. *DeGolyer Estate, courtesy of Peter Flagg Maxson.*

The lush mesquite growth around the windmill indicates ample water in the area. Ranchers welcome the vegetation and the water that makes it possible. *UTPB, Permian Historical Society.*

Lea County:
Maljamar, Jal, and Hobbs

In contrast to the social chaos of Wink, the communities developing from discoveries immediately north in Lea County experienced little tumult. The 1926 extension of the Maljamar field from Eddy County northeast into Lea County gave the latter a bountiful first field. The manager of the Maljamar Oil Company, Mel E. Baish, a former piano teacher, located the test well on a hunch. A drought had resulted in the death of cattle, and piles of bones attracted Baish's attention. Since the flat terrain dissuaded a professional geologist from recommending a drill site, Baish decided the bone pile would do and drove a stake. Maljamar Baish No. 1, completed on July 16, 1926, yielded approximately 100 barrels daily. Although the well did not signal a major discovery, it excited sufficient interest to bring several large companies into the area, including Ohio, Marland, Skelly, and Humble.

The following year a strike in the Jal field, in the southwestern corner of the county, served to link the rich production of Winkler County with the New Mexican discoveries. Texaco's Rhodes No. 1, located in unproven territory, created much interest, since it forecast an important extension of the Winkler production to the south. The Empire Gas and Fuel Company, Humble, and Magnolia, as well as Texaco, found huge quantities of natural gas along with the oil. To capitalize on this discovery, El Paso Natural Gas Company laid a sixteen-inch pipeline from Jal to El Paso, 200 miles away. Completed in June, 1929, the pipeline furnished gas for homes and factories in El Paso. Before gas entered the pipeline, it was treated at processing plants the company built at Jal. This project began the expansion of the company into an industrial leader.

Before Jal became established as a major producer of natural gas, another discovery to the north brought Lea County into the limelight as a leader in the Permian Basin. On November 8, 1928, the Midwest Refining Company's State No. 1 opened the Hobbs field, which became New Mexico's largest. Although the town of Hobbs experienced some of the social disruptions that characterized any boom town—a red-light district, speakeasies, and dance halls open twenty-four hours a day—it was not another Wink.

The fact that the boom could build on an established community tempered the social disruption. Located in a completely flat area, the Hobbs field defied the usual surface evaluations of geologists. To compensate, Midwest experimented with a magnetometer to ascertain the presence of petroleum. This machine sent electrical impulses to rock formations beneath the surface; the reflected impulses measured the density of formations and thus indicated the possibilities of finding oil. Although this magnetic probing lacked accuracy, Midwest successfully employed it with the discovery well, which happened to be on the edge of an enormously productive reef.

Lea County was the second New Mexico county to experience oil booms. The Maljamar field in the west-central part of the county—about thirty-five miles due east of Artesia—opened in July, 1926, and the Jal field in the southwestern corner of the county, in November, 1927. With the discovery of oil near Hobbs in 1928, the county became a major producer in the Permian Basin. Jack Nolan, the photographer, has described the nature of this 1930 gusher at Hobbs. *Jack Nolan Collection, PBPM.*

Soon after the Hobbs field developed, Humble established a regional office there. The warehouse is pictured in 1930. *W. L. Caruthers Collection, PBPM.*

Single employees of Humble lived in this bunkhouse in the Hobbs field in 1930. Married employees living in the Humble camp at Hobbs got not only a house for their families but also a garage for their cars. *W. L. Caruthers Collection, PBPM.*

When workers came in from the oil patch in the mid-1930's, they found downtown Hobbs bustling. *Mrs. John J. Taylor Collection, PBPM.*

The unusual horizontal nature of this night fire at a Jal well in October, 1934, was caused by gas seepage that hovered near the ground. *Frank Forsyth Collection, PBPM.*

80 / OIL IN WEST TEXAS AND NEW MEXICO

The El Paso Natural Gas Company built a plant at Jal in 1929 to process gas before sending it through a sixteen-inch pipeline two hundred miles to El Paso, where it was marketed. Witnessing the lowering of a steam still at the Jal plant in 1935 was Hugh F. Steen, *far right*. Steen later became president of the company. *Samuel D. Myres Collection, PBPM.*

These three huge wooden cooling towers were part of the soda-ash treatment of sour gas from the Jal field in 1936, a process used until 1946. *Samuel D. Myres Collection, PBPM.*

Replacing the earlier stills and cooling towers, these treating towers of the El Paso Natural Gas Company's Jal plant No. 4 removed hydrogen sulfide (the odor of rotten eggs), carbon dioxide, and water vapor from gas piped from the wells. *Texas Mid-Continent Oil & Gas Association.*

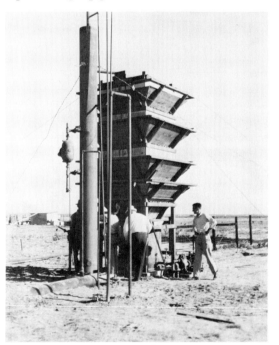 

Left: A gas-treating unit in the Monument field, about ten miles southeast of Hobbs, September 29, 1937. *Graybeal Scrapbook, PBPM*. *Right*: J. W. Graybeal, a Humble production engineer in the Hobbs District, December 21, 1937. Graybeal's extensive photographs of oil activities in the Permian Basin showed a keen sense of the industry's historical importance to the region. *Graybeal Scrapbook, PBPM.*

Left: This Pure derrick in Lea County has a monkeyboard on all four sides. The derrickman, standing on the monkeyboard, guides the ends of pipes as they are run into or out of the hole. Modern masts have a monkeyboard on only one side. *Texas Mid-Continent Oil & Gas Association. Right*: Conoco's Bell Lake No. 1, in Lea County, ignited March 16, 1954, and was extinguished sixty days later. On the right is the corrugated tin shield behind which oil well firefighters typically stood. A new president of Conoco announced after this well caught fire that the company would have no more blowouts or fires, and it did not, according to Charles D. Vertrees, a veteran Conoco geologist. The new president insisted on safety and careful attention to drilling procedures that would prevent fires. *Charles D. Vertrees Collection, PBPM.*

New Mexico oil scouts assembled in Hobbs, 1948. *Front row, left to right*: Bobby Walker, Bob Knoeffel, Hank Sweeney, Justin Newman, Ralph Hickman, and John Burton. *Center row, left to right*: Roger Harrell, Wade Smith, Bennett Anderson, Ray Eudaily, Dan Rodgers, Arnold Brown, Slick Fraser, Fred Tyner, and Hank Carhey. *Back row, left to right*: Homer Stilwell, Russell Estes, Phil Hays, Stanley Markley, E. H. Miller, Charlie Corbett, Bill Smith, Henry Littlejohn, Red Langley, Red Harrington, and Dick Whitson. At the weekly assembly, or "check," scouts exchanged information. *Raymond M. Eudaily Collection, PBPM.*

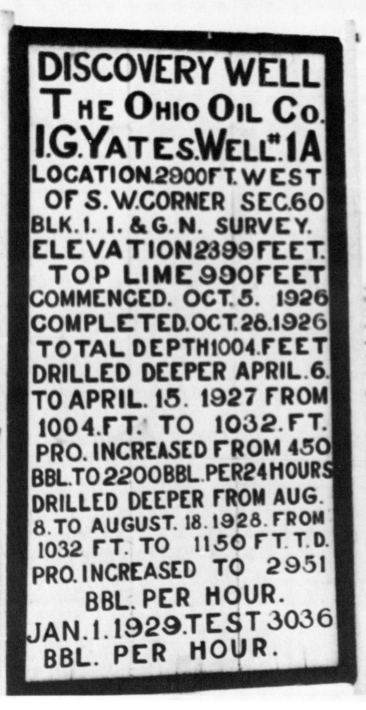

DISCOVERY WELL
THE OHIO OIL CO.
I.G.YATES.WELL.# 1A
LOCATION.2900FT.WEST
OF S.W.CORNER SEC.60
BLK. I. I. & G. N. SURVEY.
ELEVATION2399 FEET.
TOP LIME 990FEET
COMMENCED. OCT. 5. 1926
COMPLETED.OCT. 28.1926
TOTAL DEPTH1004.FEET
DRILLED DEEPER APRIL.6.
TO APRIL. 15. 1927 FROM
1004.FT. TO 1032.FT.
PRO. INCREASED FROM 450
BBL.TO 2200BBL.PER24HOURS
DRILLED DEEPER FROM AUG.
8. TO AUGUST. 18.1928. FROM
1032 FT. TO 1150 FT. T.D.
PRO. INCREASED TO 2951
BBL. PER HOUR.
JAN.1.1929.TEST 3036
BBL. PER HOUR.

Michael Benedum persuaded the Ohio Oil Company (later Marathon) to drill four test wells in exchange for partnership in the Pecos County leases. The first three wells were dusters, but the fourth, I. G. Yates No. 1A, became a prolific producer. Geological maps indicated the hilltop seen in the background as the place to drill. Frank R. Clark, the geologist in charge, argued that the structure they had mapped was large and that they were as likely to hit oil by drilling on level terrain as on the hilltop, thereby saving considerable expense. His theory proved correct, and in twenty-three days the shallow well began production. Although this marker specifies 1,004 feet, the most authoritative sources give 997 feet as the depth at which production of one hundred barrels per day began. Thereafter, as the hole was deepened, measurements accelerated to thousands of barrels per hour. *American Petroleum Institute.*

Pecos County: The Yates Giant

BEHIND the 1926 discovery of oil in Pecos County lay years of planning on the part of Ira G. Yates, on whose land the test well was drilled. Ranching on some 50,000 acres had not made Yates affluent, beset as he constantly was by drought. As early as 1915 he began leasing acreage to petroleum companies. Over the next eight years, he signed various leases that garnered him less than $10,000. At the end of 1923, he negotiated a lease with the Transcontinental Oil Company that eventually made him a multimillionaire. Transcontinental, owned by Michael Benedum, paid him $4,000 initially and $2,000 annually for drilling rights on 8,000 acres. The company's geologists then mapped the area and identified a structure where they thought oil would be found. This structure, Yates Dome, proved to be the site of the discovery well, some two miles west of the Pecos River.

Rather than bear the entire cost of wildcatting in Pecos County, Benedum made the Ohio Oil Company (parent of the Mid-Kansas Oil Company) a partner in the operation in exchange for its agreement to drill four test wells. After three dry holes, I. G. Yates No. 1A, spudded October 5, 1926, hit pay at the amazingly shallow depth of 997 feet on October 28. Drilled on a site indicating an anticline below, the well flowed 135 barrels per hour. When deepened to 1,150 feet, Yates No. 1A yielded 2,950 barrels an hour. On August 18, 1928, the well flowed 70,824 barrels—enough to be classed with the Lucas gusher that ushered in Spindletop on January 10, 1901.

The Yates discovery, thirty miles south of Rankin, quickly drew the expected hordes of "leasehounds" and those merely curious. The only accommodation possible was in rancher Yates's red barn, which he gladly converted into a rooming house and rented. That income, though, paled in comparison with the $180,000 he got from leases the day after the well came in. After the barn burned, a community of Red Barn sprang up nearby, although it lost out to a permanent town named Iraan for Ira and Ann Yates.

The prolific production of the Permian Basin fields, particularly Hendrick and Yates, far outstripped need or demand. Nor could the industry handle the glut economically. It made little sense to pump oil that had to be stored in expensive tanks, especially when its high sulfur content quickly corroded the metal tanks. W. S. Farish, president of Humble, wishing to avoid costly storage and the excess that upset the economics of the industry, called Permian Basin leaders to Houston for a conference in August, 1927. Pointing out that rapid production depleted valuable reserves and shortened the lives of fields, Farish urged producers to leave oil underground until needed. In September at a meeting in Fort Worth, important major producers agreed to the voluntary prorationing Farish suggested. On July 1, 1928, the Texas Railroad Commission began administering prorationing according to a specific code. Because of prorationing and its magnificent reserves, the Yates field has produced more crude than any other in the Permian Basin. Its recent annual output has increased to cope with the growing energy shortage, proving unmistakably the value of conservation. In 1979 Pecos County's petroleum production garnered $1,227,154,341—the first time in Texas history that a county's annual output topped $1 billion. The Texas Railroad Commission still regulates production, but, since the acute shortages of the 1970's, allowables are near the wells' maximum output.

While the Yates giant has dominated production in Pecos County, the Fort Stockton area has been the scene of exploration and production since the 1920's. Most notable was the Phillips Petroleum Company's University EE No. 1, abandoned as a duster at 25,340 feet in February, 1959. This well, costing over $2 million, held the depth record until 1970. Although a dry hole, the well was far from a total loss, since it furnished valuable information about geological structures, deep drilling techniques, and the tensile strength of materials. Over the 732 days of drilling, Phillips used 347 bits, the last of which it gold plated and put on display at its headquarters in Bartlesville, Oklahoma.

When news of the Yates discovery well spread, eager "leasehounds" converged on the Yates ranch and Yates sold $180,000 in leases on October 29, 1926. The only accommodation for those who needed to stay overnight was Yates's red barn, where the rancher installed cots and partitions. *Texas Tech, Southwest Collection*.

Although the red barn burned down in 1928, a cluster of buildings sprang up nearby. With the establishment in 1929 of a post office (the building on the right), Red Barn, Texas, was in business. In 1933 it boasted twenty-five residents. The Halamicek brothers, W. A. and Paul, owned most of the hamlet's buildings. *UTPB, Permian Historical Society*.

The village of Iraan developed four miles north of Red Barn. Named after Yates and his wife, the town's name is pronounced "Ira-an." Oil storage tanks are visible on the hill in the right background. When drilling was most active in the Yates field, Iraan's population neared 3,000, but by the late 1930's when this picture was taken, the decline had begun. Today it numbers around 1,000 residents. *UT, Archives.*

Pipes had to be hauled into the Yates field so that pipelines could transport oil to refineries. This truck had no difficulty fording the Pecos River. *Samuel D. Myres Collection, PBPM.*

The Ohio Oil Company's Yates No. 30-A, gushing 80 feet through a fifteen-inch casing, was completed September 23, 1929, at 1,070 feet. The well produced a world record of 204,682 barrels per day. *Abell-Hanger Foundation Collection, PBPM.*

Residents of this Humble Pipe Line Company camp could enjoy the wide sweep of striking terrain in the Yates field. Humble piped virtually all of its production to its refinery at Ingleside, on the north side of Corpus Christi Bay. *Texas Mid-Continent Oil & Gas Association.*

The canyons of the Yates field provided excellent collecting places for oil seeps. At the beginning of the Great Depression, thieves skimmed the oil and sold it. *Samuel D. Myres Collection, PBPM.*

Three and one-half miles from Iraan, the Soma Oil and Gas–Noelke No. 1 blew out in June, 1940, saturating the ground for half a mile around. More than 30,000 barrels of crude collected in the canyon behind the well. *Abell-Hanger Foundation Collection, PBPM.*

Crew of the Soma O. and G.-Noelke No. 1, *left to right:* Wes Pittman, Cy Pittman, Boyd Cox, Howard Coburn, W. T. Pittman, and L. G. Bloomintrip—all in oil-field clothes. Three unidentified men with hats are on the right. *Wes Pittman Collection, PBPM.*

Frank R. Clark, chief geologist for the Mid-Kansas Oil Company, a subsidiary of the Ohio Oil Company, was responsible for locating the great Yates field. Here Clark, *right*, receives the American Petroleum Institute's Certificate of Appreciation, December 11, 1957. *Abell-Hanger Foundation Collection, PBPM.*

Although the Yates field has overshadowed other fields in Pecos County, there has been oil activity around Fort Stockton since the 1920's, as this photograph of the Liberty Realty and Development Company well demonstrates. As water flows from the hole, the workers in the center, a black, a Mexican, and an Anglo, are flanked by two men who have come to watch the proceedings. Because there were few blacks in West Texas, it was uncommon to find them working on wells. *Texas Tech, Southwest Collection.*

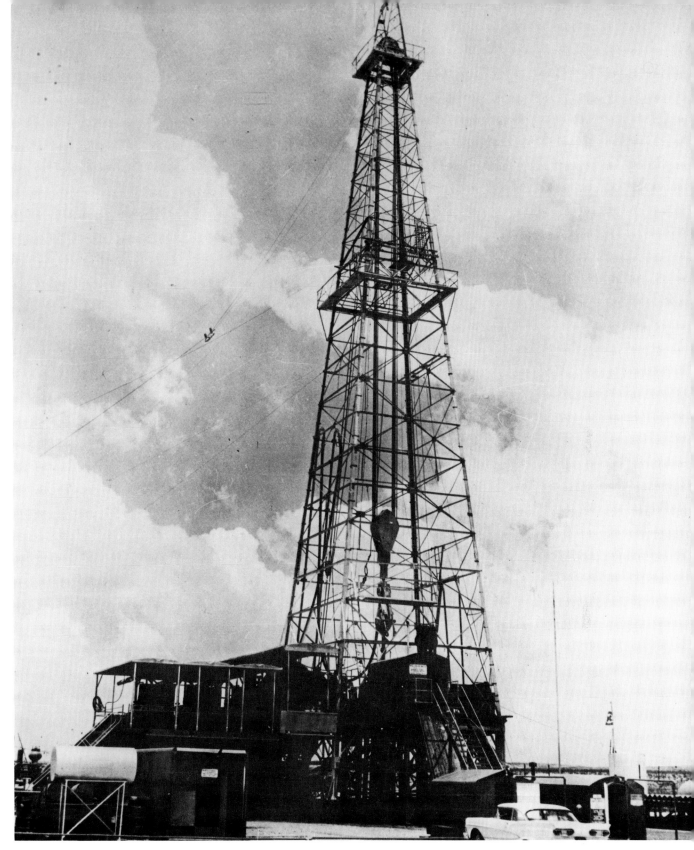

Phillips Petroleum Company's University EE No. 1, near Fort Stockton, the first well in history to go below 25,000 feet. It reached this depth in September, 1958, then was abandoned, at great financial loss. This 136-foot derrick could hoist a load of one million pounds. *UTPB, Permian Historical Society.*

A Fort Stockton tank yard, January 22, 1940. *Samuel D. Myres Collection, PBPM.*

Situated in Pecos County, about five miles west of Grandfalls near the river, these storage tanks symbolize both the continuity of oil production throughout the county and corporate change. As the fine print on the sign indicates, George T. Abell participated in developing the wells filling these Sunray DX tanks. In 1968 that company merged with its parent, Sun Oil (now the Sun Company). Currently the Cactus Operating Company of Wichita Falls controls the lease, consisting of four producing wells and three water-injection wells. *Texas Mid-Continent Oil & Gas Association.*

Discoveries in the Late 1920's: Garza, Ward, and Irion Counties

ALTHOUGH Garza County has never been a blue-ribbon producer among Permian Basin counties, its history of oil exploration involves the fascinating entrepreneur C. W. Post. Born in Illinois in 1854, Post first got to know Texas while working as a realtor in Fort Worth in the early 1890's. Through this business, he became acquainted with West Texas lands, but before he could make any investments there, he suffered a nervous breakdown and entered a sanatorium in Battle Creek, Michigan, for treatment. While recuperating, he studied W. K. Kellogg's production of health foods, especially breakfast cereals, in that city. Inspired by Kellogg, who began his food enterprises in connection with his Seventh-Day Adventist dietary beliefs, Post sought to develop a caffein-free coffee substitute. The result, Postum, made from a mixture of molasses, bran, and wheat, earned Post a handsome reward. When he introduced Post Toasties in 1906, he directly challenged Kellogg's leadership in breakfast cereals and further enlarged his fortune. Having laid the basis for what later became General Foods, Post returned his attention to West Texas.

With his income steadily swelling from nationwide sales of breakfast food and drink, he undertook a social experiment in Garza County. In 1906 he bought 213,000 acres (some in Lynn County), paying $600,000 cash. On this land he would develop a model community, centered around a new town, modestly called Post City. The town conjured up memories of a New England Puritan village, where residents were expected to work faithfully and to abjure worldly vices, especially liquor. So that his cotton farms would have a market for their produce, Post established the Postex Cotton Mills, thus ensuring economic harmony and balance for his social experiment. Among the visions that Post entertained was increasing the community's prosperity by producing oil. Despite the facts that no one had discovered oil in the Permian Basin in 1910 and the closest production was 200 miles away in Petrolia, Post financed drilling. The test well, approximately 100 feet west of the Post City boundary, extended to 1,394 feet, where the driller lost his cable tools and abandoned the well. Had the well gone deeper, it would have tapped the rich Garza pool.

The discovery of oil in Garza County came only in May, 1926, when B. V. Blackwell hit pay at 2,404 feet. While the well produced only 5 barrels the first day, it justified Post's original hopes. Subsequent developments have proved that the county's best production flows from the Garza field, which encompasses Post (as the town is now named). Post's greatest attraction currently is the fenced compound in the west edge of town where two wells produce from six different pay horizons for about 170 barrels daily. This modest production typifies Garza County's oil contribution.

While Garza County has never been in the mainstream of Permian Basin oil, either geographically or economically, Ward County, whose initial production also dates from the late 1920's, qualifies on both counts. But Ward County's prosperity from oil has not depended entirely on its own production. With the Texas and Pacific Railroad running through the county, any production in surrounding areas was likely to enhance Ward County's economy. When the Hendrick field developed in 1926, the nearest transportation facilities lay to the south in Ward County. Monahans, Pyote, and Wickett, all located on the railroad, began to prosper from Winkler County's bonanza field. Then in 1928 Shipley Hayzlett No. 1 became Ward County's first producer. The Shipley field, some thirteen miles south of Monahans, disrupted civic composure in the town as its workers roistered about.

The Great Depression stultified the oil economy of Ward County, as elsewhere, but exploration continued. In 1933 the Estes field, developed by the Humble Oil and Refining Company, became an important producer. Among those investing in this field was Sid Richardson, whose successes seemed to validate any operation in which he was involved. In the 1940's another field opened to continue Ward County's importance as a producer. Named for the county seat, the Monahans field lay just three miles northwest of the town. The Shell Oil Corporation was active in this field, drilling a series of Sealy-Smith wells, the first striking oil on August 7, 1942. This well initially flowed 2,852 barrels daily from 10,364 feet.

Realizing that the town of Monahans lay atop a potentially rich pool, the Signal Oil and Gas Company in 1968 undertook an experiment to tap the oil. Since a zoning ordinance prohibited drilling within 200 feet of any building, Signal put down a dozen wells on the south, east, and north of the town, using directional drilling techniques to reach the oil. These wells, which necessitated 950 leases from lot owners in the town, produced 410 barrels a day.

Discovery of oil in Irion County dates from 1928, as it does in Ward County. There the similarity between the two ends. In 1946 Irion County had only eight producing wells, but subsequent discoveries have increased that number. Although Irion County now has about two dozen small fields, among them Sugg, Mertzon, Barnhart, Bingham, Rocker B, Tankersley, Sixty Seven, and Lucky Mag, production through 1976 has amounted to 24,531,580 barrels—slight by Permian Basin standards. Sheep raising remains the mainstay of the county's economy.

In 1906 C. W. Post, the cereal magnate, created Post as a planned community, investing millions of dollars in its development. One of his desires for an economically self-sufficient settlement—a town surrounded by productive fields—involved petroleum production. In 1910 he drilled this well just west of the city limits. When the driller lost tools in the hole at 1,394 feet, he abandoned the well. Note the makeshift wrench poles. *Samuel D. Myres Collection, PBPM.*

Left to right: D. E. Lounsbery, George T. Abell, and Tom Allen in Post, the Garza County seat, 1925. Members of the geological crew of Midwest Exploration Company, an affiliate of Indiana Standard, these college-trained men mapped several areas within the Permian Basin. The year following this visit to Post, oil was discovered in Garza County. *Abell-Hanger Foundation Collection, PBPM.*

Today, as in July, 1975, the date of this picture, the greatest tourist attraction in Post is this well site, where pumps draw from six different pay horizons. Owned by the Bond Operating Company of Dallas, the wells are near the 1910 drilling site. *Samuel D. Myres Collection, PBPM.*

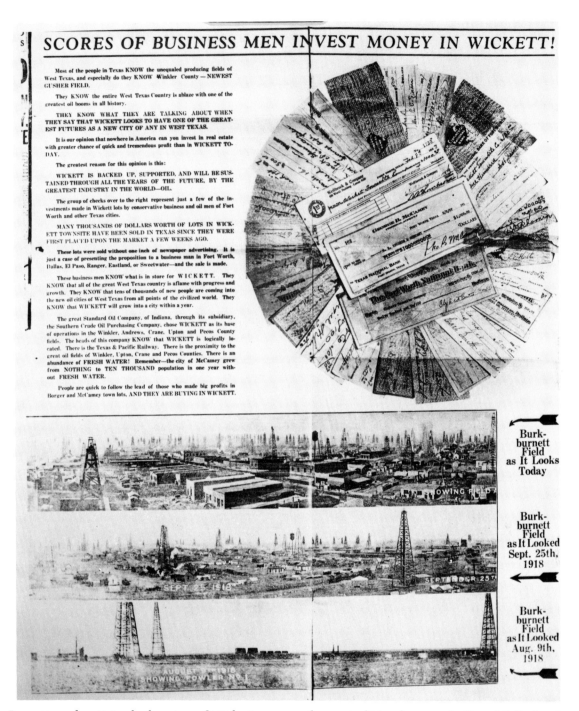

Inventive advertising by boosters of Wickett, seven miles west of Monahans on the T. and P. Railroad. Standard Oil of Indiana, through a subsidiary, Southern Crude Oil Purchasing Company, established Wickett in 1927 as a tank farm and refining center for crude from the Hendrick field to the north. Named for Fred H. Wickett, chairman of Pan American Petroleum and Transport Company, another Indiana Standard subsidiary, it never rivaled other boom towns in the region. In prominently displaying checks of Roy A. Westbrook and George B. McCamey, the Wickett Townsite Company tried to assure potential investors that oil leaders of the region had faith in Wickett's future. After three small refineries closed, the town dwindled to around 500 inhabitants. *Samuel D. Myres Collection, PBPM.*

Although Ward County's first oil production commenced in November, 1928, in the Shipley field, four miles north of Grandfalls, drilling began in the county earlier in the decade. The Malita well attracted a large crowd in the mid-1920's, but produced no oil. *Samuel D. Myres Collection, PBPM.*

Humble's district office in Wickett. This June 20, 1937, photograph has one painful reminder of the Great Depression: the sign by the door says "No Help Wanted." *Graybeal Scrapbook, PBPM.*

A tank battery in the Monahans sand hills. *Mahan & Associates, Inc.*

Left: In the late 1940's Shell drilled Sealy-Smith No. 7 in the Monahans field. The 204-foot derrick was the world's tallest at the time. The height enabled drillers to use longer pipe stands than usual—45-foot lengths, instead of the 30-foot lengths then more common. By hoisting and stacking the pipe in thribbles (lengths of threes), drillers saved considerable time and expense. This well still produces eight barrels of oil per day, along with one barrel of water. *Texas Mid-Continent Oil & Gas Association*. *Right*: The Kingswood Oil Company, which had explored unsuccessfully in Loving County in 1927, drilled Sugg No. 1 in Irion County two years later. On July 29, 1929, the well, six miles north of Mertzon, had reached 7,488 feet, and there had been several shows of oil up the hole. *E. W. Harrison Collection, PBPM*.

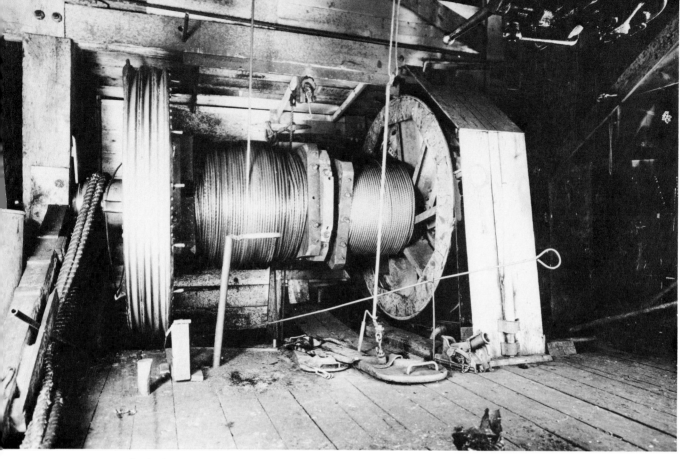

The bull wheel of Sugg No. 1 held about 14,000 feet of ⅞-inch drilling line. The hole is at the bottom of the picture, marked by "dog ears." *E. W. Harrison Collection, PBPM.*

At Sugg No. 1, *left to right*: Mr. Jones, Buda Company engine expert; E. W. Harrison, tool dresser, afternoon tour; C. L. Duncan, driller, morning tour; Jack W. Thomson, superintendent, Kingswood Oil Company; Jim McConkey, driller, afternoon tour; Tom Cartwright, tool dresser, morning tour; and H. A. Lawrence, salesman, Young Engine Corporation. *E. W. Harrison Collection, PBPM.*

Midland and Odessa

No aspect of the history of the Permian Basin has been more dramatic than the transformation of Midland and Odessa into important oil centers. Their growth documents the urban-frontier thesis that towns enhanced the development of the surrounding countryside. When the Texas and Pacific Railway built from Fort Worth to El Paso in 1881, it planted the seed for these towns along the way, and each became a county seat—Midland of Midland County and Odessa of Ector County. With the 1923 strike in the Big Lake field, the towns began to respond to the oil boom. Then with the subsequent development of fields in Eddy, Crockett, Loving, Upton, Crane, Winkler, Lea, and Pecos counties, it was obvious that Midland and Odessa lay in the middle of the vast producing area—and it became just as obvious that they would play important roles in its growth.

At the outset, there seemed nothing predestined about the role each would play. Rather, it was because of the desires and vision of certain citizens that Midland became the commercial, organizational, and banking center and that Odessa became the transportation, service, and supply center for the region. Long before the discovery of oil, Midland had developed into something of a ranching center, with pioneer rancher John Scharbauer participating in founding the town's first bank in 1888. By the time the oil boom began, Midland clearly was a city whose leaders recognized the opportunities it presented. They encouraged oil companies to locate their regional headquarters there, and many did so because of Midland's central location. Thomas S. Hogan, ex–state senator from Montana, decided that he could prosper by erecting Midland's first modern office building, and in July, 1929, he opened the twelve-story Petroleum Building. While his vision was admirable, his timing was awful, for in October the stock market crash heralded the Great Depression that left Hogan with a white elephant. The structure would later house offices of leading corporations—until they put up their own buildings—but Hogan went bankrupt.

After World War II, concomitant with the extensive development of fields north of Midland, major corporations began constructing skyscrapers that earned the city the sobriquet of "the tall city." In an act characteristic of the promotional instincts that built the community and reminiscent of Benedum's approach, some boosters brought a Texas League baseball franchise to Midland and provided the Cubs with a handsome park. Journalists have made much of the differences between Midland and Odessa—white collar versus blue collar—but the two cities function cooperatively as the urban hub of the Permian Basin. Reflecting this interdependence and mutuality, the cities jointly sponsor a symphony orchestra and chorus, whose headquarters are judiciously located at the air terminal, halfway between the two.

Odessa, never having functioned as a financial center, did not essay competition with Midland in this regard. It happily became the oil-field supply and service depot for the region, as well as a trucking center. Not only has it served as a storage point for materials shipped in from elsewhere, but it has also become a major site for fabrication of drilling equipment for use regionally and around the world. Because Odessa had established itself as a manufacturing center, it logically became the setting for the Permian Basin Oil Show, inaugurated in 1940. This biennial three-day show, which attracted more than 300,000 visitors in 1978, enables oil service and supply firms, such as Halliburton, Dresser, and the Western Company, to display their wares to the industry. Unlike Midland, which from a distance would be indistinguishable from any other prairie metropolis, Odessa can be immediately identified as an oil town because of the sizeable El Paso Products Company complex on its south side. Beginning with a natural gas plant in 1957, the complex developed into a petrochemical center including the General Tire and Rubber and the Rexene plants. Both Odessa and Midland, in their different ways, project an air of vitality and prosperity characteristic of the petroleum industry in the 1980's.

One reason that Odessa became a service and supply center is that before 1930 it was the hub of a network of highways leading to the Crane, McCamey, Andrews, Hendrick, and Ward county fields. Furthermore, it was much closer to the oil fields than Midland, especially until 1945, when Humble's Buchanan No. 1 established production in Midland County. As early as 1929 the Penwell field qualified Ector County as a major producer. The discovery well, R. R. Penn Kloh-Rumsey No. 1, began flowing on October 7, 1929, from 3,720 feet. The output was so great that temporary tanks, holding 9,500 barrels, quickly filled. The Gulf Pipe Line Company came to the rescue early in November with a two-inch line that tied the well into its trunk line from the Hendrick field. This small line typified emergency transportation measures operators took throughout the Permian Basin when wells yielded more than could be stored at the well sites. Then the Depression wreaked havoc with the economy of Odessa, because the reduction in the demand for petroleum adversely affected the demand for the services and supplies Odessa offered. The opulent Penwell field was producing oil that had little market, so the need for further drilling declined drastically.

Named Midland because of its location midway between Fort Worth and El Paso on the Texas and Pacific Railway (visible *lower right*), the city in 1928 was just developing as the commercial and financial center of the Permian Basin. *Raymond M. Eudaily Collection, PBPM.*

Midland in 1977, with the T. and P. Railway tracks in the foreground. With the erection of the spired Petroleum Building (*right center*) in 1929, Midland became the corporate headquarters for the region. Now known as "the tall city," Midland boasts offices of virtually every major oil corporation in America, plus those of influential independent producers. *Samuel D. Myres Collection, PBPM.*

While oil was not discovered in Midland County until 1945, as early as 1928 the Magnolia Petroleum Company had established a tank farm two miles east of Midland. On May 17, 1928, lightning ignited two of these 80,000-barrel tanks. The men on each side of the windmill indicate the gigantic scale of the billowing black smoke. *Clinton Myrick Collection, PBPM.*

The first female landman hired by Texaco, Becky Fisher negotiated a lease with Palmer Willis on his ranch ten miles southeast of Midland in 1975. Like the typical landman, she did the paperwork on the hood of her pickup. *Texaco Archives.*

In 1967, 529 civic and industrial leaders of the Southwest chartered the Permian Basin Petroleum Museum, Library and Hall of Fame, which opened September 13, 1975. Its exhibits document and explain petroleum oil exploration and production in the region. The three modern pump jacks in front of the museum are not, unfortunately, connected to producing wells. *PBPM.*

Behind the Permian Basin Petroleum Museum is this collection of antique drilling rigs. *Texaco Archives.*

President Gerald R. Ford spoke at the dedication of the museum on September 13, 1975. Here he views a collection of core samples. *Left to right*: unidentified Secret Service man; Emil C. Rassman, chairman, Museum Board of Executors; Congressman Richard C. White, El Paso; President Ford; Congressman George H. Mahon, Lubbock; and Russell J. Ramsland, president, Museum Board of Trustees. *PBPM.*

After World War II George H. W. Bush came to Midland to get established in the oil business. After he co-founded the Zapata Petroleum Corporation there, he moved to Houston and was elected to Congress. In 1970 he returned to Midland and nearby oil fields in his campaign as a Republican candidate for the U.S. Senate. *George W. Bush.*

Twenty miles southwest of Midland, Odessa also flourished as an urban center in the middle of the Permian Basin. As Odessa developed, it served as the trucking, supply, and service center of the industry. This 1928 aerial view gave few indications of the metropolis to come. *Jack Nolan Collection, PBPM.*

Jack Nolan, the premier photographer of the early days of Permian Basin oil, around 1930 foresaw economic development moving to Odessa's south side. Travelers could stay at the Sunshine Tourist Camp on the left between the Pickering Lumber Company and the South Side Garage. Motorists could fill their tanks at the garage or at the service station across the street. *Jack Nolan Collection, PBPM.*

In 1934 the Continental Supply Company opened its Odessa store. The spools of cable on the loading dock, as well as the lower line on the sign "Youngstown Tubular Goods," advertise oil-field supplies. *Continental Supply Company Collection, PBPM.*

The Permian Basin Oil Show, held in Odessa since 1940, attracted 80,000 visitors in 1974, as the clogged parking lots testify. The three-day show gives oil-well machinery and supply firms an opportunity to exhibit their wares. *Samuel D. Myres Collection, PBPM.*

Odessa's principal industry is its petrochemical complex, established in 1957. The most important component of the complex is the El Paso Natural Gas Company plant, shown here with its Christmas greeting in 1960. *Mahan & Associates, Inc.*

In November, 1966, the El Paso Natural Gas Company became the El Paso Products Company. This 1975 aerial view shows only part of its petrochemical complex, just south of Interstate 20. Since this industrial heart of Odessa is south of the city, Jack Nolan's 1930 prediction came true. *El Paso Products Company.*

An original component of the petrochemical complex is the General Tire and Rubber plant, which manufactures synthetic rubber. Pictured are the monomer-handling areas. *Texas Mid-Continent Oil & Gas Association.*

Near Odessa, Amoco Production Company, formerly Stanolind Oil and Gas Company (Standard Oil of Indiana), operates the North Cowden gasoline plant for a group of twenty-one owners. Here a worker turns a discharge valve on a cooling-tower pump. *Texas Mid-Continent Oil & Gas Association.*

Most Permian Basin crude has a high sulfur content. The North Cowden gasoline plant's sulfur recovery unit extracts the sulfur and processes it for marketing. *Texas Mid-Continent Oil & Gas Association.*

Four twenty-six-inch pipelines from the North Cowden field bring casinghead gas into this inlet scrubber. The North Cowden gasoline plant processes casinghead gas to recover its liquid constituents: propane, butane, and natural gasoline. The remaining "dry" gas is compressed and injected back into the producing formations to maintain pressure on the reservoir to aid further pumping of crude. *Texas Mid-Continent Oil & Gas Association.*

Odessa companies manufacture oil-well equipment for use around the world. Shown here in 1979 are two rigs in the yard of the Oil Industry Manufacturing Enterprises, just east of Odessa. *Texaco Archives.*

Ector and the Northern Counties:
Andrews, Gaines, and Yoakum

THE Permian Basin suffered a twofold blow in the early 1930's. The Depression affected the entire economy, and then the giant East Texas field, discovered by C. M. ("Dad") Joiner on September 5, 1930, compounded the problem. East Texas oil was more plentiful and cheaper than that from the Permian Basin, which reeled from the competition. Odessa lost about half its population and businesses foundered. Yet the purpose and instincts of oil men have been to find oil, and they continued this quest, hoping that by the time of discovery demand for crude would have resumed. Their quests led to two more important fields in Ector County in the 1930's: the Harper field in 1933 and the Goldsmith field in 1935. The demand for petroleum accelerated with American involvement in World War II, and the TXL field in northwestern Ector County began making its contribution late in 1944. The TXL field, taking its name from the stock symbol of the Texas Pacific Land Trust, which owned the land, exemplified the production from deep zones that became common in the region after the war. The discovery well produced from 8,950 feet, and other wells went below 9,000 feet. Because of their depth and increased expenses in the 1940's, these wells cost between $125,000 and $200,000 to drill. The Yarborough and Allen field, opened in 1947, continued adding to the prosperity of Ector County.

Three Ector County fields spawned typical boom towns: Penwell, Goldsmith, and the TXL's Notrees. Penwell, which surged to a population of 3,000 at the height of the field's production, declined rapidly afterwards. It revived briefly with new drilling in 1938 but could not survive. Now, although the village is still on the map, only used storage tanks and the remnants of a big machine shop mark its former existence. Goldsmith and Notrees have fared somewhat better, although the population of each can be measured in the hundreds. Contributing to the continuance of Goldsmith has been the presence of small El Paso Natural Gas and Phillips plants nearby. Even in its diminished state, Goldsmith maintains some of the appearance of the roistering boom town of the late 1930's. Perhaps the most distinctive feature of Notrees is its descriptive name. Like most boom towns, Notrees has a business district consisting of a few adjoining wooden buildings that house such enterprises as a drugstore, grocery store, beer joint, cafe, and barbershop.

After production began in Ector County in 1929, most of the further Permian Basin discoveries lay in counties directly to the north: Andrews, Gaines, and Yoakum. Of particular significance in Andrews County, whose discovery well was Deep-Rock's C. E. Ogden No. 1, completed December 18, 1929, was the fact that many of the producing wells were on University Lands, thus adding to the endowment. In the 1930's five new fields, the Fuhrman, Means, Parker, Emma, and Shafter Lake, established Andrews County as a major producer, and in the 1940's the Fullerton, Embar, Midland Farms, Mabee, and Dollarhide fields continued the pattern. The last, showing that oil fields did not respect geographical boundaries, extended into Lea County, New Mexico. Located near the county seat, the Andrews field became one of the county's finest, opening on June 6, 1953, with Humble's State University "O" No. 2, producing 230 barrels daily from 11,050 feet. These fields have combined to rank the county fifth in production in the Permian Basin. In 1974 the county produced 56,167,108 barrels of crude, plus about 50 billion cubic feet of gas. Andrews, the county seat, enjoyed steady and orderly growth as a result of the oil boom, stretching as it did over three decades.

The last classic boom town in the Permian Basin developed with the discovery of the Wasson field, straddling the border of Gaines and Yoakum counties. Discovered by C. J. ("Red") Davidson, a veteran driller from Fort Worth, the Wasson field's first well (in Yoakum County) showed oil at 5,085 feet on September 28, 1935. The second well, financed by Amon G. Carter, publisher of the Fort Worth *Star-Telegram*, and the Continental Oil Company, which had absorbed Marland and Texon Oil and Land, was A. L. Wasson No. 1, completed in June, 1937. These wells, plus many more, became part of the prolific Wasson field. As the field began to develop, two promoters founded Wasson in 1935 at the northern edge of Gaines County. When it appeared that the major production would be to the north, one of the promoters decided the town should be transplanted nearer the wells. Since the Denver Producing and Refining Company had opened the new pool in Yoakum County, the new town would be Denver City. To persuade citizens to move, the promoters donated land to the Methodist Church. And to minimize protest by means of legal injunctions, the promoters chose a Sunday morning in November, 1939, to transport buildings from Wasson to Denver City. A little gunfire failed to dissuade those behind the move, and Denver City took root. A bulldozer cleared away sagebrush and mesquite to create a main street, along which businesses sprang up. Like Andrews, Denver City grew in an orderly fashion, with few rowdy elements in the population that reached 5,000 in 1941.

Although other counties in the northern tier of the Permian Basin, such as Hockley and Cochran, did not experience the dramatic strikes that resulted in a Denver City, they nonetheless have produced oil in continuingly important quantities. Hockley County combined the powerful economic themes of ranching and oil when the Texas

Company brought in a well on the old Slaughter ranch in 1937. The Texas No. 1 Bob Slaughter (Guerry) began flowing 502 barrels a day from 4,443 feet after three treatments with acid, opening the enormous production of the Slaughter field. Eight years later, just to the north of this vast field, the Levelland field also began, with Texaco's No. 1 L. Y. P. Montgomery-Davies. By the end of 1974, this field, like the Slaughter field lying across both Hockley and Cochran counties, had yielded 228,698,521 barrels from an average well depth of 4,900 feet. The Levelland field, measuring approximately eight miles wide by fifteen miles long (with a long east-west axis), is only slightly smaller than the Slaughter, which it resembles in dimensions.

A central separator and metering station in the Levelland field, which opened in January, 1945. The field is just north of the Slaughter field, and both of them extend eastward into Cochran County. Levelland, originally known as Hockley City, was also created by C. W. Post. *Texas Mid-Continent Oil & Gas Association.*

Left: Ector County has been blessed with oil riches. Odessa, the county seat, developed into a major service and supply, as well as petrochemical, center. On October 7, 1929, R. R. Penn Kloh-Rumsey No. 1 established the county as an important producer. Townspeople quickly gathered to see the new derrick, around which the town of Penwell developed, just below the Concho Bluffs west of Odessa. *UTPB, Permian Historical Society. Right*: Jack Nolan used his photograph of the ignition of Skelly-Amerada University No. 1 on April 27, 1930, as an advertisement for his studio. Located in the Penwell field, the well exploded because of a 360-quart nitro shot. *Jack Nolan Collection, PBPM.*

In the Penwell field, Humble employees are using the tuboscope to apply corrosion-preventing coating inside the pipes, September 2, 1942. The tuboscope, which was fairly new then, is also an inspection device. *Graybeal Scrapbook, PBPM.*

Left: Humble drilled these Penwell field wells in a straight line, as this September, 1945, photograph shows. Such spacing characterized much of the drilling in the Permian Basin. *Texas Tech, Southwest Collection*. *Right*: Affirmative action in the Penwell field: Barbara Lou Odom working as a roustabout in 1979. *Texaco Archives*.

The Penwell field has proved to be long-lived. In 1976 Robert J. Trammel, a Texaco employee, made a welding repair on a horsehead pump. The power converter on the truck enabled him to do welding in the field. *Texaco Archives*.

Broderick and Calvert's No. 1 Elliott F. Cowden opened the Harper field in Ector County on September 23, 1933. Standing in front of the slush pit are, *left to right*: two unidentified; Harry L. Smith, driller; Fred King, tool dresser; unidentified; John H. Black, Dallas; Glenn Black, Midland; Al Broderick, Fort Worth; Joe Kelly, Midland; O. C. Harper, Midland; George Calvert, Fort Worth; W. A. Black, Black and Sons, Contractor; and Elliott Cowden, landowner. *Abell-Hanger Foundation Collection, PBPM.*

The Goldsmith field began with a gas well in 1934 and followed with an oil well the next year. The town began as Ector City in May, 1937, the date of this picture, but became Goldsmith in June. The main thoroughfare, Scharbauer Street, took its name from the prominent Midland family that had extensive ranching land in the region. As the derricks in the background indicate, the village developed hard by the field. *UTPB, Permian Historical Society.*

The Goldsmith plant of the El Paso Natural Gas Company, about three miles west of Goldsmith. The plant, *center foreground*, gets residue gas from the Phillips Petroleum Company's unit toward the highway. The El Paso plant removes impurities and compresses the gas before transmitting it through a pipeline. *Texas Mid-Continent Oil & Gas Association.*

In late 1941 Charley Judkins No. 3, a mile and a half north of Odessa, came in as one of the Permian Basin's largest producers. In this picture the Texas Railroad Commission is conducting a six-hour test of the well's potential. Developers of the well are C. O. Davis, L. A. Beecherl, and C. A. McDaniel, *first, second, and fourth from left.* The two pipes on the left are flowing into the tank, and those on the right are pumping oil out as fast as it flows in, indicating that the well yielded 550 barrels an hour. *Jack Nolan Collection, PBPM.*

Located in the extreme southwest portion of Ector County, the Yarborough and Allen field began producing in September, 1947. Humble riggers have this steel derrick about halfway up. *Texas Mid-Continent Oil & Gas Association.*

Robert A. Estes took this picture of buffalo grazing around Humble No. 1–Yarborough and Allen, which opened the field, producing from the Ellenburger formation at 10,500 feet. At the time, Estes was the assistant division civil engineer for Humble's West Texas Division. *Robert A. Estes.*

The Cities Service Oil Company developed the TXL field in the northwest portion of Ector County at the end of 1944. In September, 1945, night drilling illuminated the sky for miles around. *Texas Tech, Southwest Collection.*

C. R. Smith, Cities Service West Texas district production superintendent, and Davis S. Bodie, the TXL farm boss, in front of a horsehead pump. *Texas Tech, Southwest Collection.*

Notrees, Texas, sprang up after Cities Service opened the TXL field. Twenty-five companies drilled around 300 wells in an area nine miles long by two miles wide. The wells produced from five horizons. This May, 1949, photograph shows business booming in the treeless settlement. *Texas Tech, Southwest Collection.*

Just north of Ector County, Andrews County experienced an oil boom in 1930. Andrews, beginning as a one-street town, resembled most other boom towns in the Permian Basin. Conoco's familiar triangle sign denoted a filling station, which along with the cafe, grocery store, and drugstore made up downtown Andrews. *Samuel D. Myres Collection, PBPM.*

Left: Andrews County's Boner No. B-2 took only two months to drill, marking its completion with this blow-out on March 3, 1935. *Frank Forsyth Collection, PBPM*. *Right*: This 1943 photograph shows an Andrews County field whose wells are spaced with geometric precision. The prescribed spacing was one well per forty acres, but wells could be drilled closer together if they produced from different pay horizons, as was the case here. *Muldrow Aerial Survey Collection, PBPM*.

Standing at the base of R. M. Means No. 1 in Andrews County, in 1965, are, *left to right*: J. W. ("Ham") Pinnell, B. F. Seay, J. B. McNeil, and C. A. Farmer. In May, 1965, Andrews County produced its billionth barrel of crude. *Samuel D. Myres Collection, PBPM*.

Stanolind's South Fullerton gasoline plant at Andrews. *Texas Mid-Continent Oil & Gas Association.*

Left: In the foreground of this Superior Oil Company unit on University Lands in Andrews County, this heater-treater separates water and gas from crude so it can be pumped through a pipeline. *Mahan & Associates, Inc. Right:* A water injection well in the Mabee field in southeastern Andrews County, 1979. *Texaco Archives.*

By far the most important economic development in Yoakum County has been the development of the Wasson field, which began in 1936. Responsible for that development was C. J. ("Red") Davidson, sitting here at the typewriter in his Fort Worth office. His three companions are unidentified. Whereas T. G. Hendrick kept a Fort Worth newspaper under his cuspidor, Davidson used the more conventional rubber mat. *Abell-Hanger Foundation Collection, PBPM.*

Left: Yoakum County's Shell Dowden No. 1, spudded April 15 and completed May 11, 1938. *Charles D. Vertrees Collection, PBPM. Right*: The Slaughter field, taking its name from the huge C. C. Slaughter ranch, began in Hockley County in 1937. This well in the Slaughter field produces from two pay zones. *Texas Mid-Continent Oil & Gas Association.*

Postwar Booms: Canyon Reef and Spraberry

AFTER World War II, two significant finds confirmed continuing vitality in the Permian Basin: discoveries of the Canyon Reef in Scurry County in 1948 and the Spraberry trend in 1949, running through Dawson, Martin, Midland, Glasscock, Upton, Borden, and Reagan counties. Scurry County had had production since J. J. Moore No. 1 began in October, 1923, but the Canyon Reef discovery was of an entirely different magnitude. About eight miles north of Snyder, the county seat, Standard Oil Company of Texas Brown No. 1 came in November 19, 1948, and 1949 saw genuine boom conditions in the field, which measured thirteen miles laterally and ten miles vertically. Early in 1950, 490 wells were flowing, with another 187 being drilled. Most of this enormous production was routed to refineries through pipelines, but four natural gasoline plants were constructed. Texas Railroad Commissioner Ernest O. Thompson, recognizing the import of the Canyon Reef field, called it "the outstanding oil event in Texas in 1949."

The commission was sufficiently concerned with the rapid production of crude to institute well allowables of 112 barrels per day in April, 1950. As it investigated the situation further, the commission realized that gas pressure in the field had diminished alarmingly. It therefore recommended that operators unitize the field to institute strict conservation measures to ensure optimal recovery. With 80 percent of the royalty interests participating, the Scurry Area Canyon Reef Operators Committee (SACROC) formed in 1951 and undertook water flooding of an area including 47,000 acres with 1,229 active wells. Fifty-three water-injection wells pumped eight million gallons underground daily, the water coming from Lake Thomas in the southwest corner of the county. In 1968 Chevron, a subsidiary of Standard of California, began injecting the field with carbon dioxide from forty-six gas-injection wells. This $175 million experiment greatly enhanced the recovery of crude. In 1973 Chevron's SACROC No. 4 pumped the one billionth barrel from the Canyon Reef field, and in 1974 Scurry County's 94,172,788 barrels led the Permian Basin.

This extended activity in Scurry County, which attracted such investors as Bob Hope and Bing Crosby, inevitably affected Snyder. Its population within two years of the Canyon Reef discovery leapt from 4,200 to 20,000. This increase in population not only taxed the town's facilities, it overran them. People slept wherever they could—mostly in house trailers, but some under bridges and in wrecked and abandoned automobiles. Between 1949 and 1950 one bank's deposits skyrocketed from $2 million to $16 million. Despite these boom conditions, Snyder managed to assimilate prosperity without degenerating. Its institutions, including the county government, were sufficiently mature to enable it to accommodate reasonably well to the influx of people and wealth—monumental traffic jams and crowded cafes notwithstanding. Also, by 1948 oil fields no longer attracted the riffraff of former days, so Snyder did not have to contend with major lawlessness.

The Spraberry trend vied with Scurry County in the late 1940's and the 1950's for the Permian Basin's spotlight. In 1951, 810 wells, or 18 percent of those drilled in the Basin, tried to hit the trend. This drilling defined the immense dimensions of this geological phenomenon known as a stratigraphic trap. About sixty miles long, the Spraberry was twenty-five to thirty miles wide in Reagan and Upton counties and ten to fifteen miles wide in Glasscock and Midland counties. By 1953 rapid flowing from the Spraberry wells had produced a sharp drop in oil production, along with a great increase in gas production. In the process, reservoir pressures lessened markedly. Having no facilities for gathering and treating the gas, producers flared it in the field. To prevent this waste, the Texas Railroad Commission closed all Spraberry production from April 1 to December 1, 1953, when facilities were ready to handle the natural gas. This was the first instance of the Commission's closing a large field to prevent the waste of gas.

In addition to the continued development of the Canyon Reef field and the Spraberry trend in the early 1950's, that decade and the two following saw further exploration in the Permian Basin. Since the major fields had been identified, most of the drilling sought to mark the limits of those fields. The 1960's, in particular, saw increased emphasis on the search for natural gas, and Ward County became a leader in gas production. With the Middle Eastern oil embargo that began in 1973, the economics of domestic petroleum production underwent a dramatic change. It became profitable to rework old wells, to use secondary and tertiary recovery methods, to drill existing wells deeper in search of more crude, and to produce from wells that would formerly have been only marginal. Doubtless, much petroleum remains to be found in the Permian Basin, and, though the discoveries will likely be modest by historical standards, they should be economically worthwhile because of the energy shortage. Likewise, much petroleum remains to be produced from existing fields, thus confirming oil's preeminence in the Permian Basin well into the future.

W. W. Lechner and E. I. Thompson organized the Lou-Tex Corporation, which began oil exploration in Scurry County. In 1922 Mrs. Lechner drove the location stake for Lou-Tex's J. J. Moore No. 1, located in farmer Moore's cornfield. *W. W. Lechner Collection, PBPM.*

Riggers setting timbers for J. J. Moore No. 1, the Scurry County discovery well. *W. W. Lechner Collection, PBPM.*

Left: The J. J. Moore No. 1 derrick under construction. Production began October 9, 1923. The well has produced more than half a million barrels. *W. W. Lechner Collection, PBPM. Right*: A bit of oil-field humor: W. W. Lechner at J. J. Moore No. 1. *W. W. Lechner Collection, PBPM.*

The real boom in Scurry County began in November, 1948, with the first strike in the Canyon Reef field, about eight miles north of Snyder. McCormick No. 5 was one of the Canyon Reef's prolific wells. *Scurry County Museum.*

August 28, 1930.
Well No. 1 - Murffee
180 Quart Shot
Owen & Sloan Oil Co.
Snyder, Texas

The Owen and Sloan Oil Company's Murffee No. 1, near Snyder, responds to a shot of 180 quarts of nitroglycerine on August 28, 1930. *Mrs. George L. Wright Collection, PBPM.*

Left: While drilling was shut down on Fuller No. 19 in Scurry County, a scout took a nap. Oil scouts, employed by major corporations, reported on wells being drilled. They traveled long distances to gather information. *UT, Archives. Right*: This jackknife rig is drilling in the Canyon Reef field next to a chicken yard. The hen seems undisturbed. The jackknife mast is designed so that it can be erected for drilling, then lowered and moved to a new well location. *Texas Mid-Continent Oil & Gas Association.*

The immediate area around wells can become saturated with crude oil, as in this Scurry County scene in 1951. Environmental concerns in recent years have led to cleaner operations. *UT, Archives.*

On October 8, 1973, Chevron Oil Company's SACROC No. 4 produced the Canyon Reef formation's billionth barrel of oil, pictured on the left. While the county's early production in the 1920's was respectable, the Canyon Reef field made Scurry one of the nation's most prolific counties. This 85,000-acre field contained reserves estimated at four billion barrels. The well, the billionth barrel, and the Texas Historical Commission's marker are on U.S. Highway 84, just north of Snyder. *Samuel D. Myres Collection, PBPM.*

The Sun Oil refinery, Snyder, July, 1975. *Samuel D. Myres Collection, PBPM.*

Development of the Spraberry trend, beginning in 1949, was a major boon. Concentrated in Midland, Glasscock, Upton, and Reagan counties, the trend also led to discoveries in Martin, Dawson, and Borden counties. Night drilling in the Spraberry trend transformed derricks into striking sentinels of light. *Texas Mid-Continent Oil & Gas Association.*

The Science of Exploration

In the early days of the petroleum industry, many oil men scoffed at the application of scientific knowledge to the search for oil. Practical experience in the oil field, they thought, was far more useful than learning derived from books and laboratories. Such attitudes probably stemmed from two sources: the actual successes that unlettered men had had in discovering oil, beginning with Colonel Edwin Drake at Titusville, Pennsylvania, and the anti-intellectualism that has characterized much of American history. Without doubt, empirical methods—along with hunches—had yielded important results. Those who inferred underground oil from surface seepages often found what they sought. Then there were those intuitionists like Pattillo Higgins, who from the early 1890's kept insisting on the existence of a huge pool beneath a slight mound outside Beaumont, but who could find no one to listen to him. His stubborn persistence finally paid off when he persuaded an experienced mining engineer, Captain Anthony F. Lucas, to drill beneath the salt dome that produced the surface configuration at Spindletop—with the known results of January 10, 1901.

By the turn of the century, however, the work of university-trained geologists was commanding increasing respect among the oil fraternity. It stood to reason that those who made a scientific study of subsurface regions would be more likely to predict the occurrence of petroleum than those without such training, so oil companies began to employ geologists. Although Mitchell County's discovery well in the Permian Basin resulted from an educated guess, the genuine significance of the Permian Basin oil boom, as documented by Santa Rita No. 1, rested on explicitly scientific foundations. As mentioned earlier, Johan A. Udden, chief of the University of Texas' Bureau of Economic Geology, had carefully studied university-owned lands in West Texas and in 1916 issued a report and map showing where oil and other minerals were likely to be found. The Big Lake field dramatically demonstrated the utility of geologists for the oil industry, although their record from that time forward did not lack imperfections. In another part of Texas, for instance, throughout the 1920's major oil companies had had their geologists in

Rusk, Gregg, and Smith counties, but they all reported negative findings. A dogged wildcatter, C. M. ("Dad") Joiner, proved them all wrong in 1930 by discovering the East Texas field, the world's greatest at that time.

In the Permian Basin geologists took inspiration from the results of Udden's pioneering work. During the 1920's and afterwards, they trekked across the arid deserts with their surveying equipment tethered on the outside of their cars. In addition to their measuring devices, they invariably carried firearms and wore high leather boots to protect themselves from the region's reptiles. They even on occasion had snake-bite medicine in case the guns and boots failed. Field geologists gathered samples of rocks and took those of potential interest to laboratories for further study. Based on their findings, companies could make reasonable decisions about drilling.

Another scientific aspect of oil exploration involved seismography. Adapting techniques of studying earthquakes, seismographers (or seismologists) induced subsurface explosions and recorded the resultant sound waves on graphs. Petroleum-bearing formations produced distinctive waves that enabled seismographers to interpret the findings with a considerable degree of precision. The work of these scientists removed a large amount of guesswork from exploration, as the extensive Permian Basin fields now document. But few seasoned oil men deny that luck may still play an important role as they undertake the discovery of oil. As Sid Richardson stated in the November 30, 1954, issue of *Look*: "Luck helped me every day of my life. And I'd rather be lucky than smart, 'cause a lot of smart people ain't eatin' regular."

In the nineteenth century predicting the presence of oil underground was the province of sooth-sayers, doodlebugs, divining rods, and various other unscientific methods. Sometimes educated guesses about the earth's topography yielded amazing results, as at Spindletop in 1901. As petroleum exploration moved into the twentieth century, however, trained geologists became more integral to the scheme. The geology department at the University of Texas took a special interest in the University Lands in the Permian Basin, as Dr. Udden's 1916 report demonstrated. Among the geologists who roamed over much of West Texas searching for signs of oil were Berte R. Haigh and Gentry Kidd, shown here on the Miller brothers' ranch in Culberson County, 1926. On the running board is the tripod for a plane table, and a stadia rod rests on the rear fender. Geologists used these instruments in surveying and mapping surface structures. *Berte R. Haigh Collection, PBPM.*

A geological surveying crew's alidade sits on a plane table. Used in field mapping, the alidade is a sighting device that can measure both horizontal and vertical distance. Such measurement of surface structures is geologists' oldest field-mapping technique. The center truck's tower enables seismologists to insert explosive devices into the earth. *Texas Mid-Continent Oil & Gas Association.*

Left: Geologists scrutinize rock samples for evidence of petroleum. Ralph Hawkins, head of a Shell three-man field team, uses a hand lens to study a shale fragment. *Shell Oil Company. Right*: Field teams send likely samples to a laboratory, such as this one in Snyder, where paleontologists examine them with more elaborate equipment, November, 1949. Paleontologists search for minute fossils that indicate the nature and age of a particular formation. *Texas Mid-Continent Oil & Gas Association.*

Left: Major oil corporations, such as Shell, maintain geological archives. These well specimens are from bit cuttings and core samples. *Shell Oil Company. Right*: A Schlumberger man shows the electric well log he has just made. His sophisticated equipment is housed in the truck on the right. The well log, or "Schlumberger" (pronounced slumber-jay) interprets the type, porosity, permeability, and saturation of a formation, just as an EKG reveals the action of the human heart. *Texas Mid-Continent Oil & Gas Association.*

This Shell cartographer in Midland is transferring data from well logs onto a map. The map's projection of contour lines makes possible interpretation of the structure of the underground formation and hence of its oil-bearing potential. These well logs show with different colors the proportionate depth of different types of rock, such as limestone, sand, dolomite, or shale. On the right are the geologist's and paleontologist's notations. *Shell Oil Company.*

Among the contributions of science to oil exploration, seismography has been notable. The seismograph is a portable instrument that registers vibrations set off by dynamite charges in the earth. When the sound waves hit potential oil-bearing formations, they reflect on the seismograph, where a ray of light records the sounds' impressions on sensitive paper. Preliminary to the work of the seismograph crew is the geological mapping of an area. In May, 1949, a Cities Service seismographic crew surveyed the U–Bar Ranch near Portales, New Mexico—the far northwest corner of the Permian Basin. *Texas Tech, Southwest Collection.*

A Shell seismograph crew working in the Permian Basin in an earlier day. The man in the cab is on the "shot point" radio, informing other members of the crew that detonation is imminent. *William L. Buckley Collection, PBPM.*

Left: The detonation of the charge placed 150 feet down resulted in this geyser of rock, mud, and water. *Texas Mid-Continent Oil & Gas Association. Right*: This land shot of 750 pounds of dynamite produced a formidable eruption. *William L. Buckley Collection, PBPM.*

Stringing the wire from the truck, where the seismograph was located, to the explosive charges. *Texas Mid-Continent Oil & Gas Association.*

Earl Burford, member of a seismograph crew, nonchalantly sits on one thousand pounds of dynamite. The letter accompanying this photograph remarked that Burford was "not guarding the powder, but the booze that was always stocked in the shipment." *William L. Buckley Collection, PBPM.*

Whereas some members of seismograph crews got their diversion from resting on a pile of dynamite, others enjoyed playing blackjack, as did this Shell crew in Lamesa, 1929. *William L. Buckley Collection, PBPM.*

Blackjack was too tame for the Stewart brothers, Harry and Bill, whose recreation in the Lamesa camp was boxing. *William L. Buckley Collection, PBPM.*

The Technology of Production

OIL-WELL drilling in the Permian Basin has reflected a wide range of technology and technological changes. This technology relates to the two methods of "making hole": cable tool and rotary drilling. The method used depends on a variety of circumstances, such as terrain, the nature of the formation being drilled, the expected depth of the hole, the equipment available, and the source of power. In some places one type of equipment is essential, and in others either could be used. A general impression is that cable tools are the more primitive and that they have largely been supplanted by rotary equipment. As with all generalizations, this one has some validity, yet in the Permian Basin both methods are currently used.

The earliest wells in the Permian Basin, including Post's unsuccessful 1910 venture and T. and P.–Abrams No. 1, the region's discovery well, employed cable tools, which operated on the percussion principle. Sharp edges of a heavy bit, attached to the end of a cable, gouged a hole in the earth. As the bull wheel lowered and raised the bit repeatedly, its spinning motion, plus the weight, chewed into the ground. The driller ran water into the hole to soften the earth, and at certain intervals he used the bailer to remove the sludge produced from the water and cuttings. Advantages of cable-tool drilling are that the equipment is simpler than rotary tools and therefore more economical. Also, since it is slower, there is less chance of drilling through a pay formation. This slowness may be a disadvantage, along with the fact that some formations (such as sand) render cable tools useless. Cable tools are good up to about four thousand feet, but would be impossible to use for appreciably greater depths.

As their name indicates, rotary tools perform by drilling in a circular motion, and power to turn the rotary has ranged from the lone mule of Corsicana's 1894 discovery well to gas and diesel engines capable of turning a bit at 30,000 feet. The basic components of modern rotary equipment are a rotary table, located on the derrick floor, and the drill stem with a bit on the end. The bit may be either immobile or mobile. An immobile bit accomplishes its purpose through simple rotary action—as any circular drill would. A mobile one, called a roller bit, was developed by Howard R. Hughes in 1909. This early roller bit had two cones with unevenly spaced teeth. Rolling with the drill's circular motion, the hard steel cones ate into the hard rock. Subsequent improvements have greatly enhanced the capabilities of roller bits. A great advantage of rotary bits is that their holes are rounder and straighter than those drilled with cable tools.

Because oil must be sought in many unhandy places, such as the mountainous Yates field, the transportation of drilling equipment has presented challenges. From the early days, portable rigs for both cable tools and rotary equipment have enabled drillers to get in places where traditional derricks were impractical. Even such derricks, plus other heavy oil-field equipment, have often been moved intact. The technology of production through the years has accurately reflected existing levels of mechanical understanding and capabilities—technology that embraced designing derricks, lubricating bits, fracturing oil-bearing formations, pumping and storing oil, and injecting water and gas back into the formations to enhance secondary recovery.

While geologists did much to reduce the element of chance in oil exploration, their work held no guarantees. Old-timers, who scoffed at the new-fangled scientific ways of predicting the presence of oil, said that the certainty of oil could be documented only by Dr. Drill. Early drilling in the Permian Basin was by the cable-tool, or percussion, method. As the heavy bit, suspended from a cable, pounded into the earth repeatedly, it bored a hole. H. A. ("Red") Coulter sits on a drilling stool at a Winkler County well. The three-tiered stool enabled Coulter, who began on the top step, to turn the temper screw (at the base of the worm gear) from a comfortable position. By turning the screw, the driller kept the proper tension on the cable as the bit chewed deeper into the earth. *Texas Tech, Southwest Collection.*

After repeated pounding into the earth, the bit lost its cutting edges. Before the bit could be dressed to restore the edges, it had to be brought to a white heat in the forge. Ford Chapman took this picture in 1937 when he worked as a driller for Sid Richardson. *Ford Chapman.*

The tool dressers placed the heated bit on the anvil and reshaped its cutting edges with sledge hammers. *Abell-Hanger Foundation Collection, PBPM.*

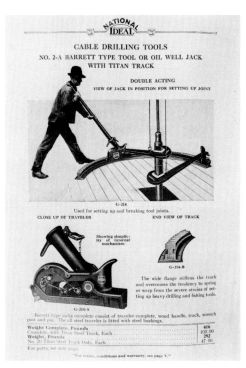

Left: To soften the formation and to lubricate the bit, water is run into the hole. When the cuttings, water, and sludge become too thick for the bit to have proper effect, the driller bails the hole with the long cylinder pictured above the hole. After the filled bailer is brought to the surface, the driller empties it into a dump box next to the well, from which the mud and cuttings run into the slush pit. The bailer in the previous photograph is in the emptying position. *Texas Tech, Southwest Collection.* *Right*: On the cable-tool derrick floor, the titan track lies near the hole. The oil well jack or Barrett-type tool travels on the track to screw and unscrew joints of pipe, since the wrenches are too heavy to be handled manually. The small cables extending upward are attached to wrench poles high on the derrick. *Abell-Hanger Foundation Collection, PBPM.*

A drilling crew perched on the mast of an antique portable drilling rig. *William L. Buckley Collection, PBPM.*

A cable-tool National spudding machine drilling M. A. Smith No. A-15 in the Yates field, September 1, 1937. *Graybeal Scrapbook, PBPM.*

Left: This National No. 3 cable-tool outfit has been inexpensively modified into a rotary rig powered by a steam boiler and engine. George T. Abell used this equipment in the Netterville field in Pecos County in the mid-1930's. *Abell-Hanger Foundation Collection, PBPM. Right*: Suspended from the crown block at the top of the derrick are the traveling block that raises and lowers the drill pipe and casing, the hook, the swivel with the mud hose entering it, and the square kelly at the bottom. Kellys, either square or hexagonal, are hollow tubes through which drilling mud is pumped to the bit. A bushing secures the kelly, which then attaches to the drill pipe. *Texas Mid-Continent Oil & Gas Association*.

While cable tools began drilling in the Permian Basin, rotary rigs soon followed. By the time rotaries appeared in the region, they consisted of fairly complicated machinery, in contrast to the simple mule-powered rotary that put down Corsicana's first well in 1894. This 1950's scene on a Sinclair rig shows a typical rotary operation: roustabouts using heavy wrenches to set up a pipe joint. The drawworks are directly behind the drill stem. *Texas Mid-Continent Oil & Gas Association*.

The Technology of Production / 151

Left: The round rotary table turns the entire drill stem. On Ohio-McDonald No. 5 in the South Eunice field in New Mexico, March, 1937, the kelly drive, manufactured by the Pressure Drilling Company, is raised above the rotary table. *Graybeal Scrapbook, PBPM.* *Right*: A Guiberson type-C drilling head on Gulf-Daugherty No. 31 in the Kermit field, March, 1937. The drilling head is in the cellar under the rotary table. *Graybeal Scrapbook, PBPM.*

The most characteristic rotary bit is a roller bit, such as this twenty-seven-inch model. Such bits can be as small as six inches in diameter. *American Petroleum Institute.*

Left: A diamond core bit being washed above the rotary table. *UT, Archives.* *Right*: A Magobar Company employee checking the viscosity of mud in a mobile testing unit. *Mahan & Associates, Inc.*

These mud or slush pumps on Annie Armstrong No. 2 in the Wasson field circulate mud from the pit to the kelly hose, August 1, 1940. Hoists maneuver the suction pipes around the pit. *Graybeal Scrapbook, PBPM.*

TOOLPUSHER ON A ROTARY RIG

BY
Slim Willet

Although Slim's lyrics incorporate some of the terminology of rotary drilling, they reveal more about the life-style of this particular tool pusher. The tool pusher is a drilling company's top man on a rig. *Mrs. Jimmie Moore.*

Left: Just as the terminology of drilling has been incorporated in song lyrics, so it has also been the subject of cartoons, as in this one, circa the 1920's. Scouts provided the "dope" on wells. *Mrs. Roy F. Gardner Collection, PBPM. Right*: At every turn, oilmen have employed technology to facilitate the discovery of petroleum. The ingeniousness of engineers and mechanics has harnessed machines to the heaviest demands of the oil field. Because assembling a rig from scratch was expensive and time-consuming, the Penrod Drilling Company employed Rumbaugh, Inc., to skid its rig from one Gulf drilling site to another, October 7, 1956. *Abell-Hanger Foundation Collection, PBPM.*

Left: This portable rig, including the drawworks and engine, is mounted on the bed of a truck. Note how the guy wires attach to the front of the truck, making the unit self-contained. *American Petroleum Institute*. *Right*: Practically anything in the oil field could be skidded, for example, this tank and separator used in the acid treating and testing of C. Scharbauer No. B-6 in the Goldsmith field, February 9, 1938. *Graybeal Scrapbook, PBPM*.

John Marks's Caterpillar and trucks moving an engine, engine house, and substructure to J. B. Tubb No. B-1 in the Sand Hills field near McCamey, April 4, 1938. *Graybeal Scrapbook, PBPM*.

The finger board on the mast of W. A. McCutcheon No. 10 in the Kermit field. This derrickman racked pipe stands into the finger board as they were pulled out of the hole on January 11, 1940. *Graybeal Scrapbook, PBPM.*

These air drilling compressors furnish air for "dusting" in lieu of mud on this rotary rig. *Mahan & Associates, Inc.*

A Halliburton "frac" job in progress. Hydraulic fracturing consists of injecting gel water and sand into the gas- or oil-bearing formation at high pressures that force the strata apart, thus promoting a freer flow of oil or gas. The fluid is stored in cylindrical tanks, and the trucks carry engines that force the fluid into the formation. *Mahan & Associates, Inc.*

In the Brown-Altman pool, about two miles south of Kermit, J. B. Walton No. B-1 was decorated in 1938 with this Christmas tree—an assemblage of control valves, pressure gauges, and chokes—mounted on the casinghead to regulate the flow after well completion. *Graybeal Scrapbook, PBPM*.

This band wheel pumps fourteen shallow wells, as power from the wheel activates the pump jacks. Once common, these central power plants have been largely supplanted as small electric motors and diesel engines have become practical for individual wells. The long jack rods required constant maintenance. *Texas Tech, Southwest Collection*.

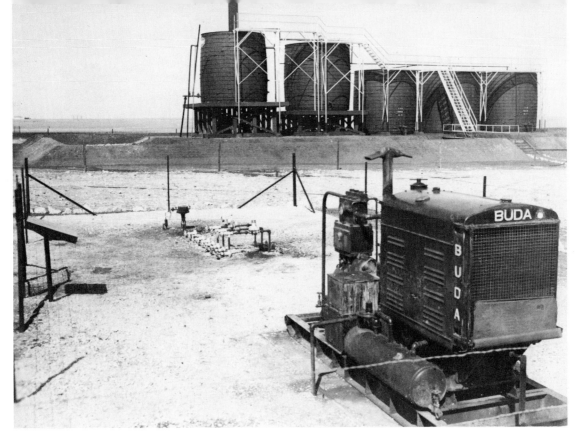

In the Vacuum field, in Lea County, the Buda engine provided power for pipelines. Between the engine and Humble tank battery is a manifold for controlling pipeline flow. On the left, as water settles in the gun-barrel tank, the water is drained, while clean oil flows into the other tanks, March 5, 1941. *Graybeal Scrapbook, PBPM.*

To promote secondary recovery of petroleum, this gas-injection facility operates in the Anton-Irish field in Hale County—at the northern extreme of the Permian Basin. The plant processes the gas before piping it back into the ground to maintain reservoir pressure. *Texas Mid-Continent Oil & Gas Association.*

Transportation:
From Wells to Consumers

FROM the beginning of America's petroleum industry in the northwestern corner of Pennsylvania in 1859, transportation of crude oil has presented a major challenge. Then, as later, men produced oil where they found it, irrespective of the distance from consumers. The economics of the industry has always indicated the preferability of transporting raw materials to refineries near the point of consumption. So long-range transportation has been needed for crude oil, and short-range, for products shipped from refineries to markets.

The immensity of the Permian Basin has accentuated the transportation problem, especially in the earlier days, because of the sparseness of population. Comparatively little of the petroleum produced in the Basin has been refined and consumed there, so the main challenge has been to ship raw materials to distant refineries. This shipment has been by pipelines, railroad tank cars, and truck, with the first two dominating in the transport of crude. Trucks have carried most of the refined products to service stations and other marketing outlets.

Geographic conditions have dictated the importance of pipelines in transporting Permian Basin oil. With few people, the region had only three railroads when production began in 1920. Luckily, the first major field, Big Lake, bordered on one of them. But when subsequent strikes occurred distant from any transportation facilities, it obviously was cheaper to construct a pipeline for the oil than to lay track into the area. Appropriate to the modest beginnings in the Basin, Mitchell County's first pipeline was two inches. Reflecting the growth of the industry in the region, in 1964 the Northern Natural Gas Company built a twenty-six-inch pipeline to transport gas from its field near Kermit.

As the photographs show, the construction of pipelines from the Permian Basin to points north and east involved men working in vast, and often lonely, places. The vistas were sometimes broken by mountains and rivers, but mostly the pipe trenches crossed parched deserts. When welded pipe joints replaced the earlier screwed ones, the work

From the beginning of the oil industry in the Permian Basin, pipelines have provided a major portion of the transportation of petroleum to refineries. Work on a pipeline crew—or pipelining—is demanding and probably less desirable than that in the oil field or refinery. Pipeliners are always on the move, so a man has to have some gypsy blood to enjoy the work. The first stage of pipeline work is surveying and staking out the trench for the pipe, as this Humble crew did in 1932. *Humble Pipe Line Company Collection, PBPM.*

became even more unpleasant in the summer, as the heat from welding augmented that of the season. Another aspect of pipelining is the great technological advances that have been made. Machines for digging trenches and laying pipe have become more complex. The pipe-laying machine that insulates the pipe immediately before laying it shows a high degree of technological capability. Once pipelines are operative, they require immense power to move the crude—power furnished by huge engines at pumping stations. Such engines underline the interconnectedness of heavy industry in our economy.

Those producers with wells near railroads, such as those at Big Lake and Mc-Camey, had fewer problems than those who had to lay pipelines. The number of tank cars could be adjusted to meet transportation needs at any given time. A familiar sight in the early decades of Basin production was scores of tank cars waiting on a siding to be loaded or trailing a puffing locomotive.

Trucks, whose size and sleekness have increased with the years, deliver refined products to the marketing point. At service stations, formerly and less pretentiously called filling stations, the cycle that began with geologists' explorations ends when consumers buy refined products. In the 1920's many of the Permian Basin's filling stations were fairly simple. Even today, in some out-of-the-way places one can find stations that are better on filling than on mechanical services. Yet in the region's towns and cities, modern stations usually offer complete automotive service as well as fuel. Their prices for gasoline, however, will never again approach those on the pumps of the old filling stations—prices from the age of cheap petroleum that is gone forever.

In the early 1930's ditching machines like this one dug the trenches. *American Petroleum Institute.*

By 1950 equipment was somewhat more advanced. This backhoe is removing blasted rock from a pipe-line ditch. *Humble Pipe Line Company Collection, PBPM.*

As this eighteen-inch Humble pipeline headed east out of the Permian Basin, it crossed the Llano River near London, in northeast Kimble County. Two wagon drills excavated the underwater crossing. *Exxon Corporation.*

Once the trench was prepared, pipe was laid beside it, awaiting interment. This pipe stretches out into the distance of New Mexico. *Mrs. J. A. McVean Collection, PBPM.*

After being welded and before being laid in the ground, the pipe is wrapped to protect it against soil corrosion. Blacks usually found it easier to get a pipeline job than one in the oil patch. *Shell Oil Company.*

These El Paso Natural Gas Company men are inspecting the tar coating with a holiday detector. The spring encircling the pipe has an electrical charge, and a spark jumps from the spring to the pipe when there is a hole in the coating. Such holes receive a dab of tar. *Texas Mid-Continent Oil & Gas Association.*

Left: Caterpillar tractors laying Shell's 493-mile pipeline from Jal to Cushing, Oklahoma. This Basin System line reached Cushing via Wink, Midland, and Wichita Falls, carrying approximately 165,000 barrels daily, beginning January, 1949. *Shell Oil Company. Right*: Deep in the trench, a worker pours tar onto the pipe. *Mahan & Associates, Inc.*

Tremendous power is required to move oil through pipelines. In 1939 these engines in the Iraan pumping station generated such power. *Abell-Hanger Foundation Collection, PBPM.*

A pipeline manifold in Iraan. The valves controlled all oil flowing in and out of the station. *Abell-Hanger Foundation Collection, PBPM.*

A tank farm manifold in Crane. *Abell-Hanger Foundation Collection, PBPM.*

This booster compressor station provides additional pressure to move crude through a pipeline. *Mahan & Associates, Inc.*

The worker is inserting a pig, or cleaning device, which will move through the pipeline under pressure, removing water, rust, and other foreign matter. *Texas Mid-Continent Oil & Gas Association.*

A Humble Pipe Line Company telegraph operator at Kemper Station, six miles west of Big Lake, September 1, 1948. Kemper Station pumped much of Reagan County's oil. *Humble Pipe Line Company Collection, PBPM.*

Before the 1970's most women in the employ of oil companies worked in laboratories or as office clerks, like these Humble Pipe Line Company typists. *Texas Mid-Continent Oil & Gas Association.*

For those oil fields near established railroads, transportation of petroleum by tank cars was cheaper and more efficient than constructing a pipeline. In the flush days of the McCamey field, such sights as this were common. Unlike pipelines, tank cars transport both crude and refined petroleum. *Abell-Hanger Foundation Collection, PBPM.*

From the Cosden refinery, tank trucks carried gasoline to filling stations. *Lee Jones, Jr., Collection, PBPM.*

At the El Paso Products Company plant in Odessa, gas goes through final filters before being piped into tanker trucks. *Mahan & Associates, Inc.*

The goal of the petroleum industry has been providing products for paying customers. Exploration, production, refining, and transportation are operations that culminate in marketing. This Standard Oil filling station at Pine Springs, Culberson County, on the western edge of the Permian Basin, blended the history of the Old West with that of the twentieth century. Located at a Butterfield stage camp, the filling station symbolized the oil economy that completely revolutionized the region. The Guadalupe Mountains are in the background. *UT, Humanities Research Center.*

The primitive Chevron filling station in Orla, Reeves County, doubled as the U.S. post office. *Ford Chapman.*

Transcontinental Oil Company, a subsidiary of Ohio Oil, opened the Yates field. It marketed its products under the Marathon symbol, as in this Dallas filling station in 1930. *Marathon Oil Company.*

Corporate change has eliminated these marketing names, but they were standard for decades. Prices on the pumps recall an age that now seems far distant. *Texas Mid-Continent Oil & Gas Association.*

Bibliography

Primary Sources

American Petroleum Insitute, Washington, D.C. Public relations photographic files.

Institute of Texan Cultures, San Antonio. Photographic Collection.

Mahan & Associates, Inc., Odessa. Photographic Collection.

Odessa *American*, Odessa. Photographic Collection.

Orbeck, Betty J., Odessa. Collection, privately held.

Permian Basin Petroleum Museum, Library and Hall of Fame, Midland. Abell-Hanger Foundation Collection. Bonner, J. D., Collection. Buckley, William L., Collection. Burton, Paul J., Collection. Caruthers, W. L., Collection. Chapman, Ford, Collection. Chapman, Ford, Scrapbook. Chiles, H. Eddie, Collection. Continental Supply Company Collection. Donnelly, Richard, Collection. Eastland Oil Company Collection. Eudaily, Raymond M., Collection. Forsyth, Frank, Collection. Frank, Mrs. T. J., Collection. Gardner, Mrs. Roy, Collection. Graybeal, Joseph W., Scrapbook. Haigh, Berte R., Collection. Harrison, E. W., Collection. Humble Pipe Line Company Collection. Jones, Lee, Jr., Collection. Kimball, Willard C., Collection. Kovach, John J., Collection. Lake, Frank W., Collection. Lawless, George W., Collection. Lay, L. R. "Rudy", Collection. Lechner, W. W., Collection. McVean, Mrs. J. A., Collection. Muldrow Aerial Survey Collection. Myres, Samuel D., Collection. Myrick, Clinton, Collection. Nolan, Jack, Collection. PBPM Collection. Pittman, Wes, Collection. Sivalls, Fannie Bess, Collection. Taylor, Mrs. John J., Collection. Theis, Anton, Collection. Tunstill, Granville G., Collection. Vertrees, Charles D., Collection. Wheeler, Eleanor, Collection. Wright, Mrs. George L., Collection.

Texaco, Inc., White Plains, New York. Archives.

Texas Mid-Continent Oil & Gas Association, Dallas. Public relations photographic files.

Texas State Archives, Austin. Cook, L. L., Collection.

Texas Tech University, Lubbock, Southwest Collection. Jones, Lee, Jr., Collection. Rister, Carl Coke, Papers.

University of Texas at Austin, University Archives. Oral History of the Texas Oil Industry Collection.
———, Humanities Research Center. Adkins Collection. McGregor Collection.

University of Texas of the Permian Basin, Odessa, Permian Historical Society, Archives Photographic Collection. Fluitt, Robert L., Collection. Rice, Joe, Collection. Shoopman, Bill, Collection. Tucker, Harry L., Collection.

BOOKS

Baker, Ron. *A Primer of Oil-Well Drilling.* 4th ed. Austin: Petroleum Extension Service, University of Texas at Austin, 1979.

Boatright, Mody C. *Folklore of the Oil Industry.* Dallas: Southern Methodist University Press, 1963.

Bryant, Keith L., Jr. *Arthur E. Stilwell: Promoter with a Hunch.* Nashville: Vanderbilt University Press, 1971.

Giebelhaus, August W. *Business and Government in the Oil Industry: A Case Study of Sun Oil, 1876–1945.* Greenwich, Conn.: JAI Press, 1980.

Griffin, John Howard. *Land of the High Sky.* Midland: First National Bank of Midland, 1959.

Heiman, Monica. *Johan August Udden: A Biography.* Kerrville, Texas, 1963.

James, Marquis. *The Texaco Story: The First Fifty Years, 1902–1952.* New York: Texas Company, 1953.

King, Philip B. *Permian of West Texas and Southeast New Mexico.* Tulsa: American Association of Petroleum Geologists, 1942.

Kuhn, Paul J., comp and ed. *Delaware Basin Oil.* San Angelo, Texas: Petroleum News Company, 1959.

Langenkamp, R. D. *Handbook of Oil Industry Terms and Phrases.* 2nd ed. Tulsa: Petroleum Publishing Company, 1977.

Larson, Henrietta M., and Kenneth Wiggins Porter. *History of Humble Oil and Refining Company: A Study in Industrial Growth.* New York: Harper & Brothers, 1959.

Martin, Robert L. *The City Moves West: Economic and Industrial Growth in Central West Texas.* Austin: University of Texas Press, 1969.

Miller, Sidney L. *Tomorrow in West Texas: Economic Opportunities along the Texas and Pacific Railway.* Lubbock: Texas Tech Press, 1956.

Moore, Richard R. *West Texas after the Discovery of Oil.* Austin: Jenkins Publishing Company, 1971.

Myres, Samuel D. *The Permian Basin: Petroleum Empire of the Southwest. Era of Advancement, from the Depression to the Present.* El Paso: Permian Press, 1977.

———. *The Permian Basin: Petroleum Empire of the Southwest. Era of Discovery, from the Beginning to the Depression.* El Paso: Permian Press, 1973.

Pope, Clarence C. *An Oil Scout in the Permian Basin, 1924–1960.* El Paso: Permian Press, 1972.

Rister, Carl Coke. *Oil! Titan of the Southwest.* Norman: University of Oklahoma Press, 1949.

Rundell, Walter, Jr. *Early Texas Oil: A Photographic History, 1866–1936.* College Station: Texas A&M University Press, 1977.

Ryan, Robert H., and Leonard G. Schifrin. *Midland: The Economic Future of a Texas Oil Center.* Austin: Bureau of Business Research, University of Texas, 1959.

Schwettman, Martin W. *Santa Rita: The University of Texas Oil Discovery.* Austin: Texas State Historical Association, 1943.

Tinkle, Lon. *Mr. De: A Biography of Everette Lee DeGolyer.* Boston: Little, Brown and Co., 1970.

Vertrees, Charles D. *History of the West Texas Geological Society, 1926–1969.* [Midland: West Texas Geological Society, 1973.]

Wade, Richard C. *The Urban Frontier: The Rise of Western Cities, 1790–1830.* Cambridge: Harvard University Press, 1959.

Warner, C. A. *Texas Oil and Gas since 1543.* Houston: Gulf Publishing Company, 1939.

Williamson, Harold F.; Ralph L. Andreano; Arnold R. Daum; and Gilbert C. Klose. *The American Petroleum Industry: The Age of Energy, 1899–1959.* Evanston: Northwestern University Press, 1963.

ARTICLES AND PAMPHLETS

Gard, Wayne. *The First 100 Years of Texas Oil and Gas.* Dallas: Texas Mid-Continent Oil & Gas Association, 1966.

Haigh, Berte R. "Santa Rita, the Oil Well." *Permian Historical Annual* 17 (December, 1977): 57–67.

———. "The University of Texas and Its Land." Appendix I in Samuel D. Myres, *The Permian Basin: Petroleum Empire of the Southwest. Era of Advancement, from the Depression to the Present.* El Paso: Permian Press, 1977.

Harris, Eleanor. "The Case of the Billionaire Bachelor [Sid Richardson]." *Look*, November 30, 1954, p. 83.

Herald, Frank A., ed. *Occurrence of Oil and Gas in West Texas.* Austin: Bureau of Economic Geology, University of Texas, 1957.

Olien, Roger M. "Boom Town Business: The Permian Basin Experience." *Permian Historical Annual* 19 (December, 1979): 3–11.

Prindle, David F. "The Texas Railroad Commission and the Elimination of the Flaring of Natural Gas, 1930–1949." *Southwestern Historical Quarterly* 84 (January, 1981): 293–308.

Rundell, Walter, Jr. "Centennial Bibliography: Annotated Selections on the History of the Petroleum Industry in the United States." *Business History Review* 33 (Autumn, 1959): 429–447.

———. "Photographs as Historical Evidence: Early Texas Oil." *The American Archivist* 41 (October, 1978): 373–398.

———. "Texas Petroleum History: A Selective Annotated Bibliography." *Southwestern Historical Quarterly* 57 (October, 1963): 267–278.

Turner, Frederick Jackson. "The Significance of the Frontier in American History." In *Annual Report for the Year 1893*, American Historical Association, pp. 199–227. Washington: Government Printing Office, 1894.

Index

feed stock, 53, 54
female. *See* women
Fields, Robert, 55
filling station: in Andrews, 124; in Best, 32; in Big Spring, 52; Chevron, 171; Marathon, 172; in Odessa, 109; in Rankin, 48; Standard Oil, 171; transport to, 160, 170; types of service at, 161
finger board, 156
fire, 45, 72, 80, 83, 106
Fisher, Becky, 106
Flynn, Thomas, 35
Flynn-Welch-Yates No. 3, 34, 35, 37
Ford, Gerald R., 108
Fort Sam Houston, 18
Fort Stockton: exploration near, 86, 92; 1900 well near, 9; Phillips rig near, ii; on railroad, 5; O. W. Williams from, 17; tank yard at, 94
Fort Worth, 18, 86, 95, 116, 127; railroad from, 103
fossils, 140
fracturing ("frac job"), 146, 157
Francis, Charles I., 27
Fraser, Slick, 83
fraud, mail, 63
Friend, Frank F., 24, 27
frontier, 5; agricultural, 6; natural selection on, 64; towns on, 68; urban, 103
Fuhrman field, 116
Fuller No. 19, 134
Fullerton field, 116

Gaines County, 17, 116
gamblers, 44, 64
Gardner, Roy, 29
Gary, Reginald, 13
Garza County, 95–96, 98
Garza field, 96
Garza pool, 95
gas: casinghead, 114; engines run by, 145; in injection wells, 129, 146, 159; natural, 77; pressure of, 64; production of, 31, 130; seepage of, 80; sour, 81; treating unit for, 82
gasoline: natural, 114; plants for, 112, 113, 114, 126, 129; prices of, 161, 172; transport of, 170
gear, worm, 147
General Land Office, 18
General Oil Company, 51
General Tire and Rubber Company, 104, 112
geologists: attitudes toward, 137–138; equipment of, 41, 139, 140; evaluations by, 78; mapping by, 85; for Shell, 140; training of, 137; usefulness of, 137–138, 139; in West Texas, 139, 140
Giavochini, Al, 13
Glasscock County: exporation in, 44; Spraberry trend in, 129, 130, 136; supply center for, 52; Texas Pacific Land Trust in, 5; well in, 51
Goldsmith (town), 115, 120; plant at, 121

Goldsmith field, 115, 120, 155
Goodnight, Charles, 6
Goose Creek, 9
government, county, 130
Grandfalls, 94, 100
Grant, J. W., 63
Graybeal, J. W., 82
grazing, 6, 18
Great Depression. *See* Depression, Great
Gregg County, 138
Gregory, Nalle, 27
Griffith, W. M., 27
Guadalupe Mountains, 171
Guiberson drilling head, 152
Gulf Coast fields, 9
Gulf-Daugherty No. 31, 152
Gulf McElroy wells: No. 1, 55; No. 103, 59, 60
Gulf Oil Company: camp of, 60, 70; conservation by, 74; in Crane County fields, 57, 58, 59; drilling sites of, 154; in Hendrick field, 64
Gulf Pipe Line Company, 104
gusher: Big Lake Oil Company, 31; Dean No. 1, 46; Flynn-Welch-Yates No. 3, 35; Hobbs, 78; Lucas, 85; Santa Rita No. 1, 20; Soma Oil and Gas–Noelke No. 1, 91; Yates No. 30-A, 89

Haigh, Berte R., 17, 139
Halamicek, Paul, 87
Halamicek, W. A., 87
Hale County, 3, 159
Halliburton Company, 104, 157
Harper, O. C., 120
Harper field, 115, 120
Harrell, Roger, 83
Harrington, Red, 83
Harrison, E. W., 102
Harte, Houston, 22
Hawkins, Ralph, 140
Hays, Phil, 83
heater-treater, 126
Henderson, R. H., 61
Hendrick, Thomas G., 63, 65
Hendrick field: development of, 96; discovery of, 55, 63; drilling in, 64; flooding at, 74; highway to, 104; housing in, 69, 70; near Kermit, 72; pipeline from, 104; production in, 85; snow in, 70; tank battery at, 71; Wink near, 68
Hendrick No. 11, 71
Henson, Harvey, 69
Henson No. 11, 71
Hickman, Ralph, 83
Higgins, Pattillo, 137
highways, 104
Hill, Roy, 40
Hines, W. B., 29
Hobbs (town), 77, 78, 80, 83
Hobbs District, 82
Hobbs field, 78, 79
Hockley County, 116, 127
Hogan, Thomas S., 103
hook, 151

Hope, Bob, 130
Hopi Drilling Company, 40
hose, mud, 151
hotels, 21
housing: bunkhouse, 79; camps, 30, 70; prefab, 69; shotgun houses, 44; tents, 55
Houston, 86
Howard County, 4, 44, 52, 53
Hudspeth County, 17
Hughes, Howard R., 146
Hughes Tool Company, 59
Humble (field), 9
Humble No. 1 Yarborough and Allen, 122
Humble Oil and Refining Company: bunkhouse of, 79; camp of, 44; employees of, 49, 82; in Estes field, 96; in Hendrick field, 64; in Maljamar field, 77; offices of, 79, 100; in Penwell field, 119; pipeline of, 163; riggers of, 122; separator of, 71; surveyors for, 161; tanks of, 159; wells of, 104, 116
Humble Pipe Line Company, 90, 168, 169
hydrogen sulfide, 55, 82
Hyer, Fred, 53
Hyer-Clay No. 1, 44, 52

I. G. Yates No. 1A, 84, 85
Illinois, 21
Illinois Producers Company, 35
Illinois Producers No. 3, 34
Independent Oil and Gas Company, 70
Indiana Standard Oil Company, 98
Ingleside, 90
injection, 74, 146, 159
Iraan, 85, 88, 166
Irion County, 17, 96, 101
irrigation, 4

J. B. Tubb No. B-1, 49, 155
J. B. Walton No. B-1, 158
J. J. Moore No. 1, 131, 132
jack, oil well, 149, 158
Jackson No. 2, 37
Jal (town), 77
Jal field, 77, 78, 80, 81; Plant No. 4 in, 82
Jeff Davis County, 3
Jester, Beauford, 27
Johns, C. D., 44
Johnson and McCamey (business firm), 63–64
Johnson and McCamey Baker No. 1, 43
Johnson, D. R., 27
Johnson, E. M., 33
Johnson, J. P., 43
Johnson, Lyndon B., 73
Joiner, C. M. ("Dad"), 115, 138
Jones, Mr., 102

Kansas, 21
Kansas City, Mexico, and Orient Railway, 5, 20–22, 43
Kelley, Frank H., 16

Vane, Walter, 13
Vaughn, W. J., 11
Vertrees, Charles D., 83

W. A. McCutcheon No. 10, 156
W. H. Badgett No. 1, 15, 16
W. R. Grace nitrogen plant, 54
Wade, Richard, 6
Walker, Bobby, 83
Wallen, Len, 13
Ward Commercial Photo Company, 66
Ward County: Depression in, 96; as
    gas producer, 130; highway to, 104;
    Malita well in, 100; prosperity of,
    96; ranching in, 4; University Lands
    in, 17
warehouse, 79
Warren, Ed, 27
Wasson (town), 116
Wasson field, 116, 127, 153
water, 76; for boilers, 13; drinking, 44;
    dumped on ground, 74; and flood-
    ing, 8, 129; for locomotives, 4, 6; in
    pipelines, 168; salt, 3, 15, 31, 32; to
    soften earth, 145, 149; wells for, 9,
    20
water injection, 74, 94, 126, 129, 146
Weekley, Burton F., 55, 58
Welch, Van S., 35, 39
welding, 119
wells: gas-injection, 129; offset, 36,

57, 71; reworking of, 130; water-in-
    jection, 94, 126, 129, 146
Westbrook, Roy A., 63, 66, 99
Westbrook (town), 8, 9, 10, 12
Westbrook No. 701, 8
Western Company, 104
West Virginia, 21
westward movement, 6
Wheat, J. J., 43, 46
Wheat field, 43, 46; No. 1, 43, 46
White, Richard C., 108
Whitson, Dick, 83
whores. *See* prostitutes
Wichita Falls, 55, 94, 165
Wickett, Fred H., 99
Wickett (town), 96, 99, 100
Wickett Townsite Company, 99
wildcatters, 9
Willet, Slim, 154
Williams, O. W., 17, 44
Williams, Ted, 29
Willis, Palmer, 106
Wilshire field, 50
Windfohr Oil Company, 37
Wingfield, Ed, 37
Wink: as boom town, 64; growth of, 68;
    lawlessness in, 64–65; on pipeline,
    165; sink at, 75; teamsters in, 69
Winkler County: development of,
    103; discovery well in, 66;
    Hendrick field in, 55, 63, 65; near

Jal field, 77; Keystone field in, 73;
    park in, 47; University Lands in,
    17; well in, 147
Wink Townsite Company, 67
women, 14; clerks, 169; investors in
    Santa Rita No. 1, 19; landmen, 106;
    landowners, 44, 53; roustabouts,
    119; wives of oilmen, 13, 35, 85, 88,
    131. *See also* prostitutes
Woods, Paul M, 69
Woods, Plaza M., 69
wool industry, 22
World No. 1 Cerf Bunker Wildcat No.
    4, 42
World Oil Company, 42
World pool, 45
World Powell No. 1, 43
wrenches, 149, 151
wrench poles, 8, 97, 149

Yarborough and Allen field, 115, 122
Yates, Ira G., 85
Yates, Martin, Jr., 35, 39
Yates field: Dome in, 85; opened, 55,
    172; pipeline from, 88; production
    in, 85–86; red barn in, 87; spudder
    in, 150; terrain of, 4, 90, 146
Yates No. 30-A, 89
Yoakum County, 116, 127

Zapata Petroleum Corporation, 108